MW00990081

# YOU CAN'T JOKE ABOUT THAT

# Why Everything Is Funny, Nothing Is Sacred, and We're All in This Together

# YOU CAN'T JOKE ABOUT THAT

Kat Timpf

BROADSIDE
BOOKS

HarperCollins books may be purchased for educational, business, or sales promotional use. For information, please email the Special Markets Department at SPsales@harpercollins.com.



FIRST EDITION

Library of Congress Cataloging-in-Publication Data

Names: Timpf, Kat, author.

Title: You can't joke about that: why everything is funny, nothing is sacred, and we're all in this together / Kat Timpf.

Description: New York, NY: Broadside, [2023] | Includes index.

Identifiers: LCCN 2022048265 (print) | LCCN 2022048266 (ebook) | ISBN 9780063270428 (hardcover) | ISBN 9780063270435 (ebook)

Subjects: LCSH: Comedy—Political aspects—United States. | American wit and humor—Political aspects. | Freedom of speech—United States. | Cancel culture—United States. | Timpf, Kat. | Women comedians—United States—Biography. | Women libertarians—United States—Biography.

Classification: LCC PN1929.P65 T56 2023 (print) | LCC PN1929.P65 (ebook) | DDC 818/.602—dc23/eng/20230206

LC record available at https://lccn.loc.gov/2022048265

LC ebook record available at https://lccn.loc.gov/2022048266

23 24 25 26 27 LBC 5 4 3 2 1

*To Cheens Timpf and Joan Rivers, neither of whom can read this*

# CONTENTS

# INTRODUCTION

I officially started working for Fox News in May 2015: about six months after my mom died, three months after my grandma died, and a few days after the guy who I'd thought I was going to marry broke up with me in front of my father at Coney Island.

Then, that June, as I was leaving my apartment in Bushwick, Brooklyn, to go film my first man-on-the-street comedy video for *The Greg Gutfeld Show*, my dad called to tell me that he had found our family dog, a miniature poodle named Axel, suddenly and unexpectedly dead in a puddle of his own blood-vomit on the kitchen floor that morning.

I cried, of course. I kept crying as I sat on the train on my way to the city to be silly on camera. It was something about millennials and social media, the first of many man-on-the-street videos that I'd do in Times Square with Joanne Nosuchinsky, who was on the show with me before Tyrus. I think it's still on YouTube.

When I've told people about this time in my life, the most common response has been to ask me how I was able to do it. How could I have managed this job, where I was being paid to make people laugh, at the exact same time that my own life was so depressing?

But everyone who has ever asked me that question has had it backward. Because, for me, the opposite was true:

It wasn't that I was able to manage doing comedy *despite* my misery—it was that I was able to manage my misery *because I was doing comedy.*

Believe it or not, I wasn't able to get through that time because I just didn't really care that much about my mom or my grandma or my dog or my paralyzing loneliness. Of course I did, it was some serious shit—which is exactly why it is the kind of shit that all of us need to talk about less seriously.

So much of the way we talk about sensitive subjects is wrong. We've created the wrong rules. We purposely misread each other. We create unnecessary conflicts when we should feel like we're all in this together.

When someone says, "You can't joke about that," what they really mean is "this is a subject that makes people sad or angry." This is a book about why those subjects are actually the most important to joke about—not just for me, but for all of us.

# YOU CAN'T JOKE ABOUT THAT

CHAPTER 1

# DISCOVERING THE POWER OF COMEDY

I really got into performing stand-up comedy when I was living in Los Angeles after college. I'd done it maybe once or twice before that, but I didn't really dive into it until I needed it.

Let me explain what I mean when I say I *needed* to do stand-up.

After graduation, I wasn't like every other aimless idiot out there. I had a plan.

The problem? Seeing as I was twenty-one years old, I still *was* an idiot—and, although I may not have been an aimless one, the plan that I'd come up with just may have been one of the dumbest plans in history.

See, I had a summer internship at Fox News' Los Angeles bureau that lasted two and a half months. I had secured a housing stipend for that internship, which lasted two months. I had also been accepted into the Columbia University School of Journalism and was already enrolled for the fall 2010 semester.

So, this was the plan: First, I was going to use the housing stipend for two months to rent a room I found on Craigslist, where I'd sleep on a Coleman blow-up mattress (and be too socially terrified to ever interact with my medical-student roommates, requiring me to be very strategic about when I'd make myself my Cup O' Noodles for dinner in the kitchen). Then, during the final two weeks of the internship, after my housing stipend ran out, I would live with my college boyfriend of three years, who had gotten a job in Los Angeles in his studio apartment. Then we would break up, and I would move to New York City alone and go to Columbia.

Yeah. We planned to live together for exactly two weeks ahead of a planned breakup, and then I would leave straight from that apartment and move to the other side of the country by myself.

I'm sure I'm not spoiling anything when I tell you that things didn't work out that way. The plan *was* really stupid, so I'm sure you already figured that out on your own. (What you might not have figured, though, is that College Boyfriend is now one of my best friends, and was one of just thirty guests at my wedding to someone else.)

Anyway, after I was already enrolled in Columbia—and less than a month before the start of classes—I decided that there was no way in hell that I could ever afford to pay back an $80,000 loan on an entry-level journalist's salary. So, even though attending that exact school had always been my dream, and it was very painful to do so, I unenrolled.

I decided to, instead, keep working at Boston Market, where I'd already been employed as a cashier to make some extra money during the internship. That was one hell of a job. Part of it was cleaning the bathroom at night, and *man*, do people treat bathrooms differently when they know that they won't have to be the ones to clean them. Some of my coworkers would routinely come to work rolling on ecstasy, and I'll never forget the time that one of them forgot to cut up the chicken for the next day's shift. I had to spend that entire next day explaining that, although I *was* aware that "this is Boston Market," we did *not* have any chicken. *Might I interest you in some meat loaf or turkey instead?* To be fair, some of those guys were also very fun to play beer pong with and equally kind about letting me stay on their couch when I needed to. Still, I started looking for higher-paying work as a waitress, as well as another internship where I could keep learning broadcasting skills when the Fox one was over.

That summer, I was technically an intern for Fox Business Network—but I had also been completely obnoxious about meeting as many people as I could, and trying to learn as much as possible from all of them. At one point, I had gone into the radio office and asked the women working there if they had a phone charger, and when they said yes, I followed up by asking them if they could also teach me how to do radio. When I did, they stared blankly at me, and then at each other—because, as they'd explain to me later, no intern had ever actually gone in there and talked to them before. (They would also leave food for me at my desk. Both of them are still my friends to this day. One of them was also at my thirty-guest wedding.)

Anyway, I had made it awkward, but it paid off! They helped me out, and got me an internship at KFI Radio in Burbank, which would eventually lead to my first real broadcasting job as a traffic producer and reporter.

Throughout all of this, I was still living with the boyfriend. We had even graduated from sharing a pool raft that would deflate throughout the night to a real bed that he had bought. Sure, we were fighting a lot. The space was small, and I was feeling really alone, and he was feeling the pressure of being my entire support system. So, when he stopped inside a pet store to try to sell insurance to them (that's what he was doing for work) and saw that they were giving away a random stray kitten that the owners of the store had found alone by a dumpster on the property, he grabbed the little guy and brought him home for me as a surprise. He was in terrible shape: underweight, malnourished, and suffering from a virus and worms. The vet would later tell me that the only reason a six-week-old kitten would be left alone by his mother would be that he was so sick and weak that he was slowing down the hunt for the rest of the pack. I loved him immediately. I decided in that moment that, no matter what it was, the two of us would get through it together. I named him "Sgt. Pepper," which would eventually morph into "Pepper," which would eventually morph into "Pepperoncini Pepper," which we would eventually shorten to "Cheens," which stuck.

After about six months, though, the boyfriend's mom told him that she didn't want us living together because we were too young, and convinced him to move in with some of their family in the area without me after he decided to quit his job. In his defense, I had moved in with him completely nonconsensually in the first place by just not going anywhere after I decided to not go to Columbia. I also couldn't afford pretty much anything, and most of that burden was falling on him unexpectedly, too.

I got a horrible apartment in a horrible neighborhood where I had no Internet, TV, or lobby. The water frequently went out without warning, and you could easily break into the main entrance using a credit card. Then, the boyfriend ended our nearly four-year relationship via text. After I got that text, I demanded that we talk in person. So I drove to his brother's house—I had since gotten a car—where he was hanging out. We sat in his car outside of his brother's building, and, despite my pathetic, tearful pleas, he just broke up with me *more.* I will never forget this: him in the driver's seat, me next to him in the passenger's seat, and him telling me, "You don't know anyone else here. Maybe you should just move back home." I then drove to my diner job, where I couldn't stop sobbing during my entire waitressing shift—leaving me no choice but to explain to customers what had just happened. (The plus side? It was the most I'd ever make in tips.)

Him telling me to move home strengthened my determination to absolutely *not* do that, but at the same time, he was right about one thing: Other than a few restaurant acquaintances, those radio girls that I had just met, a girl that I had only sort of known from college, and another girl who was the sister of a friend of mine from middle school, I *didn't* know anyone else. I *didn't* really have anyone. I didn't really have much of anything at all.

So, in my mind, there was only one thing to do: Go to open mics and tell jokes about my dumpster-fire life onstage. Everything was awful, but I'll never forget how great it felt to turn my pain into jokes that made me—and other people—laugh about all of it. During the loneliest time of my life, comedy became my means of connection. It was my one refuge from hopelessness, the only thing that gave me power over the things that were making me feel so powerless.

And I absolutely *had* felt powerless. Powerless, lonely, and unbelievably exhausted. For a while, my schedule was grueling: I'd get up at 4 a.m. to drive from that trash apartment in Long Beach to the Fullerton Airport, where I'd report on the traffic from a helicopter until 9 a.m. Then I'd drive straight from there to KFI in Burbank for my internship, then from there to my closing shift at a Sherman Oaks diner that ended around 11 p.m. before finally driving the hour home to my apartment, where I'd pass out on a yoga mat. (The ex-boyfriend had taken the bed that he'd been letting me use after we broke up.)

But the nights that I didn't work at the diner? I'd be out performing stand-up comedy. It's what kept me going, because I didn't feel powerless or lonely when the audience was laughing along with me. I eventually replaced my diner job with a job at a California Pizza Kitchen much closer to my apartment, which allowed me to get a little more sleep sometimes. To be clear, the "little more sleep" was hardly a match for everything else I was up against: I will never forget, for example, the time I got scabies (probably from the bus) the same week that Cheens got fleas. No one should ever be that itchy. I lost my traffic reporting job, my car, and eventually—actually, the same week as the scabies and the fleas—my apartment, forcing me to move in with this Colombian bartender guy and his family. I had sort of been seeing him from the California Pizza Kitchen job. He was a hot, stupid, tattoo-covered "recovering" heroin addict—I say "recovering" because I'd eventually discover that he was still totally abusing his Suboxone—and there were many difficult conversations involving me having to explain to him that, just because I lived with him at his family's house, that did *not* mean that we were boyfriend and girlfriend. No, no matter how many (I would eventually find out, stolen) bracelets he gifted me. The entire arrangement was a total disaster, except for the fact that it was the only time I was ever really fluent in Spanish. (It was an immersive experience that allowed me to strengthen the skills I had learned as a Spanish literature minor, similar, I'm sure, to what those other Hillsdale College kids must have gotten by spending a semester studying in Barcelona.)

Anyway, I recently found an old video of me performing a stand-up set at a bringer show (the kind where you're allowed to perform only if you bring a certain number of people to sit in the audience) at the Belly Room in the Comedy Store around that time. It was September 2011, right before I lost the apartment. I came out on the stage: a scrawny, lost twenty-two-year-old, wearing a cheap bow in her hair that contrasted sharply with her deep, raspy voice. It was even raspier then than it is now, in fact, because I was still smoking cigarettes, which worked great for my opening joke: "*I look like a nice little girl, but I sound like somebody who invites nice little girls into his van.*"

Just like the juxtaposition between my appearance and my voice made the first joke work, the juxtaposition between my appearance and the things I would admit may have made the rest of them work. Because the stuff I was saying on that stage was, by anyone's judgment, *not* funny. It was sad! For example:

*The other day a homeless man told me I looked like Macaulay Culkin. It was a bad day.*

*Maybe people from California can help me understand. People here are on diets? Which means you have extra food that you could be eating, but you're not? May I please have the scraps? Fill out my SpongeBob arms a little bit? I couldn't afford to be bulimic.*

*Then you have those friends out here who have way more money than you do, and you try to be cool about it, but it's a little awkward, you know? Like, you go to their apartment, and there's people there dressed as butlers in the lobby? Shit. I'm lucky if people are dressed in my "lobby." And you get upstairs, and they're always apologizing, like, "Oh, sorry, it's such a mess. Such a mess in here right now. Please just move my Banana Republic clothing over to the side and have a seat on my leather couch." Then, they come to my apartment, which I insist is not a good idea . . . but they want to be cool, you know? I open the door, and I'm like: "I'm sorry about my abject poverty. Have a seat on the yoga mat I sleep on every night. My back hurts so bad."*

*The only people that I know here are my ex-boyfriend and my cat. Which puts a lot of pressure on my cat. Because I require a lot of attention. I mean, I need to stand here on a stage because everyone I know is sick of hearing me talk about myself. I'm chasing him around. I'm like, "Kitty, can we please cuddle? Kitty, what went wrong in my relationship? Oh my God, kitty, Ashley looked so fat today," and he's like—HISSSSS HISSSSS HISSSSSS!*

No, none of these were the best jokes that I've ever written. Sure, they were funnier as part of that actual performance then than they are on paper now, because that's how stand-up works. That is why, by the way, every stand-up comic ever will hate you if you're one of those people who, upon hearing someone is a comedian, replies with "You're a comedian? Tell me a joke!" Still, they were not my best, but cut me some slack. For one thing, I had just started. For another thing? Who cares; I *needed* them. I needed to write them; I needed to tell them; I needed to laugh and have other people laugh with me.

What's more, the crowd actually *did* laugh—which suggests to me that there must have been people in the crowd who needed to hear what I was saying on some level, no matter how "sad" the subject matter happened to have been. Unfortunately, a lot of the things that I joked about in that set then would probably be on the list of things you "Can't Joke About" now.

Looking back and rewatching that set more than ten years after I performed it—now that my life is far better than it was, and I am living so many of the exact things that I used to dream about in those days—I can't help but feel a little sad. Not so much for my circumstances during that time, but for how we seem to be losing the very ability for people to heal in the exact sort of way that I'd been learning how to heal throughout them. The key word there, too, *is* "learning"—no one starts doing stand-up, or any kind of comedy, without being bad at it first, and people need the freedom to be able to mess up in order to figure out what works. Actually, that's true of anything, really. What unfortunately seems to be unique to comedy, though, is the lack of allowance for making mistakes.

Now, many people believe that certain subjects are sacred. That you simply can't joke about them.

These people—often the loudest voices in the room of our society—say you can't joke about death, about trauma, about poverty or illness. Serious, dark, and difficult things must be handled carefully—and absolutely *never* joked about—because it's the moral and respectful thing to do.

It's a widely accepted standard. Some might even call it "common decency."

But me? I'd call it bullshit.

I have been through some awful stuff. If you don't believe me, just read this book, and keep in mind that there's plenty of other stuff that I didn't even include. But the thing is, nothing I've ever been through has been made easier because people insisted on speaking carefully about it. If anything, the opposite is true.

In my experience, and probably yours, our cultural expectation to speak solemnly about difficult things adds discomfort to the devastating. I would argue that societal policing of levity and humor limits our ability to heal, or worse, to make connections with one another through our shared life experiences.

Life is hard enough without having to freak out that you're talking about it wrong. Why can't we all agree to make things easier by taking that pressure off ourselves?

Now, I would say that the Left is more intolerant when it comes to speech than the Right is. It's something I've seen not only in the media, but also within my own friend group and elsewhere in my personal life. The way politics relates to all of this is certainly going to be a small part of this book, but it won't be all of it . . . and not in the way that you're used to.

For one thing, I think it's important to note that the Right is not entirely immune to claiming that certain things are grounds for cancellation, or at least "not okay" to say or to joke about. Remember when Donald Trump basically suggested it was illegal for *Saturday Night Live* to make fun of him as much as it did during his presidency? Or that full-scale right-wing meltdown when Lil Nas X launched a pair of pentagram-adorned "Satan Shoes"? Or that time Kathy Griffin did a photo shoot pretending to hold up Trump's severed head, and many conservatives called for her to be prosecuted over it?

I defended Kathy Griffin's right to publish that photo at the time, and that's not just because I enjoyed so many episodes of *My Life on the D-List*. (Even though I, uh, certainly did.) It also wasn't because I loved that image, because I actually thought it was pretty gross.

No, I pushed back against those calling for her prosecution because, to me, whether or not I personally think something is vile is totally irrelevant when it comes to my core principles. To me, it's more important to live in a culture wherein a person, any person, doesn't have to worry that his or her attempt at communication or humor will result in the complete annihilation of their entire life.

As a free speech absolutist, I make absolutely no exceptions—even when someone is absolutely brutalizing *me*.

In fact, I've had a more difficult time dealing with the tough stuff in my own life—like the untimely death of my mother—because of the exact standards that were purportedly in place to help people in tough situations like mine. It was almost as if no one could have a real conversation with me, because they were so obviously terrified of saying the wrong thing.

Death and dying, of course, rank pretty highly on our culture's unspoken "You can't joke about that" list, presumably because they're incredibly traumatic. But wouldn't it make more sense that, the more traumatic an experience is, the more you would need the healing power of laughter? To me it does, and there's research that backs me up on that. In 2011, two studies out of Stanford University showed that comedy was a more helpful tool than solemnity in helping participants deal with traumatic imagery. This shouldn't be surprising, because the benefits of laughter are well documented and understood. The Mayo Clinic, for example, claims that laughing has a whole host of physical benefits—ranging from pain relief to organ stimulation to a stronger immune system—so the last thing we should do is make people too afraid to make the jokes that can elicit it.

Additionally, multiple studies from Harvard University found that trigger warnings are, at best, useless, and might even cause further harm to people experiencing trauma. Yet people still use them, and even shame others for neglecting to do so.

We are doing ourselves a huge disservice by ignoring all of this—because candid communication and humor are more than just excellent coping mechanisms. They're also amazing tools for bringing us together.

I first got the idea for this book while I was on the phone with my dad, fresh off the emergency ileostomy surgery that I'll discuss more later. He said to me, "You're only thirty-two, but I'm having a hard time trying to think of something that you *haven't* been through."

At first I laughed, and then I said, "Well, with every tough thing you go through, you're automatically building a connection with everyone else who has gone through it, too."

I may have been on painkillers at the time, but damn was I right. Going through something difficult can be an incredibly isolating experience, but it would be far less so if we could all just talk to one another without the fear of doing it wrong. Candor and comedy really *do* connect us as humans, and it depresses me to think of how much connection we might be missing out on because people are too afraid to try.

Sadly, many of our cultural norms surrounding speech come from nothing more than a thoughtless, knee-jerk adherence to decorum, even when statistics and research prove that the opposite is true.

In this book, I'm challenging that thoughtless narrative using stories from my own life, observations from pop culture and society, and even good-old fashioned research.

Honestly, even though we are talking about levity—the stakes couldn't be higher.

Openness and humor absolutely need to break free from the constraints of fear and cultural censorship. It's so important for all of us, both individually and as a society. The darker the subject matter, the greater healing that laughter can bring, disarming the darkness and making the people who are feeling isolated by their trauma feel less alone.

The truth is, anyone who has ever said "You can't joke about that!" isn't just annoying and wrong; they're also causing real harm—robbing joy, healing, and connection from the people who need it the most.

CHAPTER 2

# INTENTION ABSOLUTELY MATTERS

Every time I see some Internet moron confidently declare that a joke was "offensive" and "not funny"—and that the fact that it was just a joke does *not* matter—I want to punch a wall. Sometimes I even want to do something more destructive, like reply and involve myself in an Internet argument with idiots, or worse, columnists for *Mother Jones*.

For one thing, the phrase "that's not funny" is always an opinion. It's not the same as stating an objective fact, like "it's snowing" or "it's cancerous" or "lacrosse is not a sport."

Note: Although lacrosse may often be confused for a sport, everyone knows that the only real sports are the ones that you can potentially make real money playing. Lacrosse is not that. It's an arena engaging in a glorified form of mass masturbation to how rich their families are—because apparently, they're rich enough to waste exorbitant amounts of time and money pursuing a worthless craft. I married a former D1 lacrosse player; I would know.

No one gets to be the arbiter of what is and isn't funny, let alone someone who is clueless enough to believe not only that such a standard exists, but also that crossing it is a transgression punishable by cancellation, regardless of the joke teller's intention.

Different people find different things funny. Sometimes the same person will find something funny at one point in their life, only to consider it offensive at another point. Humor is about as subjective as it gets! (For example, I literally cannot imagine anyone who is not a twelve-year-old boy with a prepubescent mustache watching and enjoying *Family Guy*, and I say that as someone who behaves more like a twelve-year-old boy than any other adult woman I've ever met. But, alas, Seth MacFarlane's hundreds of millions of dollars signify to me that, mathematically, my view cannot be an objective fact.)

It is wrong and misguided for people to treat jokes they don't like as irredeemable offenses—let alone as offenses so severe that motive doesn't matter.

It's still happened countless times, of course. One of those times was during a 2013 episode of *Fashion Police*, when Joan Rivers joked about the sexiness of Heidi Klum's dress by saying: "The last time a German looked this hot was when they were pushing Jews into the ovens."

Rivers faced massive backlash, including from the Anti-Defamation League.

"There are certain things about the Holocaust that should be taboo," ADL director and Holocaust survivor Abraham H. Foxman said at the time. "This is especially true for Jews, for whom the Holocaust is still a deeply painful memory. It is vulgar and offensive for anybody to use the death of six million Jews and millions of others in the Holocaust to make a joke, but this is especially true for someone who is Jewish and who proudly and publicly wears her Jewishness on her sleeve."

But Rivers refused to back down, saying, "My husband lost the majority of his family at Auschwitz, and I can assure you that I have always made it a point to remind people of the Holocaust through humor."

In other words? Rivers didn't joke about the Holocaust *despite* the fact that it was a grave subject and the joke would garner controversy and attention, but *because* it would. Her intention was not to minimize the seriousness of the Holocaust, but to remind people of exactly that.

If there's something people find "not funny!" more than the Holocaust, it's probably rape.

In 2012, that's exactly what happened to Daniel Tosh, then the host of *Tosh.0* on Comedy Central. During a set at the Laugh Factory, Tosh had apparently been telling a few rape jokes when a woman in the crowd heckled him, yelling, "Actually, rape jokes are never funny!"

Tosh then allegedly replied to her, "Wouldn't it be funny if that girl got raped by, like, five guys right now? Like right now?"

I say "allegedly," by the way, because I wasn't there. The story became national news because the woman made a Tumblr post about the incident, which then promptly went viral.

People were pissed. Worse, many were the most obnoxious kind of pissed: the kind where they see their pissed-offness as a valiant display of virtue and heroism. Lindy West wannabes of all genders rushed to their MacBooks, determined that "*I'll show him!*" by banging out think pieces about what a monster Tosh was for having made that horrible, awful, not-funny joke.

The whole thing was massive. *Entertainment Weekly* stated, "it seems Tosh firmly strayed over the line." A Twitter mob called for his firing, and Lindy West herself said that cancellation of *Tosh.0* over the "public outrage" would be an example of something "integral to freedom," basically because it would represent the will of the people being realized.

Adam Martin wrote in the *Atlantic*, "[T]he problem with Tosh's joke wasn't just that it was in bad taste, it's that it was also simply bad." Margaret Lyons wrote on Vulture that although "[t]here's no such thing as off-limits in comedy," she had never heard a funny rape joke, and that "[r]ape jokes reinforce the idea that male identity is neutral and normal, and female identity is marginal and laughable. *Terrorizing and marginalizing women is hilarious, and you just can't take a joke*." (Her emphasis, not mine.)

At some point, Tosh responded by both apologizing and defending himself, tweeting: All the out of context misquotes aside, i'd like to sincerely apologize . . . the point i was making before i was heckled is there are awful things in the world but you can still make jokes about them. #deadbabies.

The *Washington Post* summed it all up this way: "A woman who was the focus of the joke got upset, she Tumbld about it, then everyone got mad at Tosh, who then apologized, but that apology didn't quite cut it, and then everyone talked about when it's okay to make a joke about rape, and the conclusion was sometimes, but only if it's funny."

Whew. Okay.

Well, first of all, it is simply wrong to say that rape jokes are never funny just because rape isn't funny. Like, no shit rape isn't funny. That's not a hot take. The number of bloggers who actually wasted their time writing hundreds of words on that, as if it's not something everyone except the most sociopathic of sociopaths already understand, blows my mind to this day.

In his apology/defense, Tosh was absolutely correct to say that we can make jokes about the darkest things in life—be it rape or the Holocaust or anything else.

In fact, I'd take it a step further. I'd say that it isn't just that we *can* joke about life's most awful things, it's also that we *should*. Humor, after all, does have extraordinary healing potential, and the last thing we should be doing is limiting the possibilities for healing among the people who need it the most. In fact, former prisoners of war in Vietnam have reported that humor was more helpful to them than religion during their captivity.

Now, to be clear, I don't think that Tosh's rape joke was among those with healing power, or any power at all, really. Hell, it didn't even make me laugh. What's more, I can actually understand why the woman in question didn't enjoy it. (She also shouldn't have been surprised that she didn't enjoy it. After all, you can pretty much guarantee that you're going to wind up upset if you heckle—the comic is almost always going to get the better of you, because that's part of the job.)

So, sure, I can see not liking the joke—but what I can't understand is the fire-and-brimstone reaction just because you didn't like it.

To everyone saying that the joke wasn't expertly crafted: Duh. It wasn't crafted at all! It was a last-minute, off-the-cuff comment in response to a heckler. Coming up with jokes on the fly is a necessary part of comedy. No, they won't always hit, but when they do? Nothing can match the energy of a moment quite like a joke that was conceived during the same one. Some of my best jokes—both on *Gutfeld* and onstage—have been off-the-cuff remarks, things that popped into my brain and flew out of my mouth before I had the chance to examine their potential implications.

The flip side, of course, is that such spontaneity also presents a risk—which is why what was perhaps one of my *worst* jokes came about the same way.

On a live episode of *The Greg Gutfeld Show* in 2020, when we were discussing Jimmy Kimmel taking a hiatus from his show to spend more time with his family, I made an off-the-cuff joke involving the difficulty of medicating a feral cat compared to medicating a human child. I'd given my cat, Cheens, his heart medication right before coming to the studio and the memory of having to hold him down by his scruff was fresh in my brain. Here's the excruciating bit.

> **ME:** I have to give my cat heart medicine every night. That's pretty hard.
>
> **GUTFELD (IMMEDIATELY REALIZING WHAT IS HAPPENING):** OH NO!
>
> **ME:** Sometimes kids need medicine, but it's a lot more difficult to hold down a cat than a kid. Babies don't have claws. If they do, you should see a priest.
>
> **GUTFELD:** Kat, I just want to remind you that Jimmy Kimmel has a son with a very serious heart condition.
>
> **ME:** Oh. I didn't know that.

I was beyond mortified. I was completely paralyzed by my own humiliation, unable to do anything but blurt out "sorry." Gutfeld and the other guests were laughing, because it was like watching someone fall into an open manhole.

I hadn't meant to hurt anyone. I thought I was just making a funny joke that could be relatable to fellow childless people living in a culture that hails parenthood as *the* catch-all excuse for getting out of things. Actually, before I said it, I'd even recalled that the article sent with the show's rundown had quoted Kimmel as saying that he and his family were healthy! I had no idea about his son, or that my comments could have been interpreted as a cavalierly cruel mockery of serious pain.

I apologized fully in the next segment the *second* I had a chance to speak. I felt so terrible that I spent the rest of the show fighting back tears, and trying to delay my anxiety from reaching full-blown panic while I still had to be on the air. Thankfully, everyone who had actually watched the entire thing was pretty understanding.

Many could tell that I had clearly been crying in the breaks and even sent along their support.

Unfortunately, it didn't end there.

The next day, someone posted the clip without context, along with the claim that I had purposely, intentionally been ripping on Kimmel *because of* his child's heart problem. Media organizations took that angle and ran with it, completely leaving out the fact that I'd simply made a mistake, or that I had apologized for how it may have sounded at the very first chance I got to do so. (Ironically? All of this happened while I was taking Cheens to the vet.)

It was a bad week. An Internet mob brutalized me; I spent a lot of time in bed crying, and Jimmy Kimmel's sister told me to kill myself. Actually, her exact phrasing was: "I extend a sincere FUCK OFF AND DIE." All I meant to do was make people laugh with a joke that I thought was harmless and maybe relatable to cat owners and the childless, but what felt like everyone thought I was a horrible, heartless bitch who didn't deserve to live. (I *loved*, by the way, when the clip later resurfaced on Reddit, and I got to go through all of this again!)

Now, if I had truly come out and said, "Do you know what? *Fuck* kids with heart issues *and* the parents who raise them!" then that kind of reaction might have been deserved. Completely and absolutely! But how does it make sense for me to have gotten a reaction appropriate to *that* when *that* wasn't what happened?

Put simply: My intention should have mattered to the mob, but it didn't.

The obvious issue with not allowing comedians to try jokes that may miss is that doing so is literally what comedy is. Anyone who has ever done stand-up, or even made a joke among friends at a party, will tell you that the only way to truly know if a joke is going to work is to try it. There is absolutely no other way, and so it's inevitable that, at some point, you're going to find yourself with an Internet mob calling for your head (or your buddy Dustin's girlfriend calling for you to never be invited over again, depending on the situation).

In explaining why someone should be canceled, people love to say that it's not that they mind offensive jokes—it *is* a comedian's job to push boundaries, after all—those jokes just need to be funny! These people clearly think that they're being reasonable, but their view demonstrates an absurd misunderstanding of how comedy works.

A better standard would be intention. If someone says something that offends us—but their intention was humor—then we should respond far differently to that than we would if the person intended to be cruel.

For example: In 2019, Donald McNeil Jr., a *New York Times* journalist with more than four decades of newsroom experience, was having dinner with some American high school students in Peru as a representative of the *Times*. During the dinner, one of the students asked him if he thought that a classmate of hers should have been suspended for having used the n-word in a video that she made when she was twelve years old. Then, as McNeil puts it, "I asked if she had called someone else the slur or whether she was rapping or quoting a book title. In asking the question, I used the slur itself."

Parents complained about the incident, and at first, an internal *Times* investigation found that McNeil should be punished for using the word, but not fired—because "it did not appear to me that his intentions were hateful or malicious," according to an internal memo written by executive editor Dean Baquet.

The week after this investigation closed, however, 150 *Times* staffers claimed in a group letter that a re-investigation was necessary. McNeil's intentions were "irrelevant," they claimed, because "what matters is how an act makes the victims feel."

*Times* leadership responded with a letter of its own stating that they "largely agree with the message." Just two days after that group letter, Baquet and managing editor Joe Kahn released a memo announcing that McNeil would be leaving the paper, stating, "We do not tolerate racist language, regardless of intent."

Unfortunately, this is not the only example of this kind of ridiculous, misguided thinking. In April 2021, *Jeopardy!* contestant Kelly Donohue lifted up three fingers on camera. Donohue claimed he had done so to signal that he had won three times; some believed the gesture looked like one that's sometimes used by white supremacists. Then 450 of the show's former contestants signed a letter demanding that the game show take action against Donohue's racist "messaging"—whether Donohue had done so "intentionally or unintentionally."

But, again, intent *does* matter. Of course it does, or we wouldn't have separate charges for vehicular manslaughter and first-degree murder. Of *course* it makes a difference whether or not Donohue meant to signal that he had won three games, or to signal "WHITE POWER!" Any rational person should be able to understand this, yet "intent doesn't matter" is repeatedly parroted anyway. It's thoughtless and illogical, but that hasn't stopped it from becoming popular enough to make people too afraid to speak.

It's especially infuriating to hear "intent doesn't matter" when it comes to comedy. If someone tries to make a joke, but ends up offending people, how should it *not* matter that that person's intention was to spread joy through laughter?

I'm not, of course, saying that you're not allowed to be upset by a joke. It's normal and healthy to have feelings, and it's healthy to express them. In doing so, though, you should remember the intention of the person who told it. Was this person trying to be hurtful or to perpetuate a harmful stereotype? Or was this person simply trying to be funny or to add some levity to a painfully tough topic?

Was this person's comment a serious expression of their deeply held beliefs or was it an off-the-cuff joke in response to a heckler?

If someone plans and then commits a brutal murder, do we treat that person the same as we would treat someone who, say, was trying to cheer up his neighborhood with a trombone solo on his balcony, but then accidentally dropped it, striking and killing a pedestrian?

Why should it be so different with jokes, then? Certainly it can't be because uttering an errant word is more consequential than ending a person's life.

Unfortunately, though, declaring intention irrelevant when it comes to speech has become common. In a piece for the *Wall Street Journal*, psychologist Paul Bloom explored this phenomenon, beginning the column with several examples of people who had been fired from their jobs over offensive comments, even though all of them said that they hadn't meant to hurt anyone. Among them was McNeil, that *New York Times* journalist who was fired for quoting a racial slur in response to a student's question about that slur after 150 of his colleagues signed a letter explicitly stating that his intentions in using the word were "irrelevant."

It continues to shock me that 150 people could have signed something so clearly absurd. If his intention was indeed "irrelevant," as they said that it was, then that would automatically mean it would have been no worse for him to have used a racial slur because he was a racist—purposefully aiming to hurt and demean black people because he hated black people. Taken to its logical conclusion, this purportedly "woke" letter actually diminishes the seriousness of racism, not the opposite.

In his column, Bloom examines the role that intention plays in our society and legal system—it matters, he points out, when it comes to a person killing another, or even spilling coffee on a laptop, but doesn't matter when it comes to things like speeding tickets. When it comes to speech, he says that most people weigh intention "for people we care about," while "zero-tolerance is something we reserve for strangers and enemies, either personal or political"—reaching the conclusion that considering intention amounts to exercising kindness:

> *And so, in the end, the argument for caring about intention is an argument for charity—for treating a stranger or even an enemy like someone we care about. It is possible that even here outcome will trump intent, particularly if someone is guilty of a string of past offenses. But charity should incline us to be more willing to take other considerations into account. And there might even be some selfish advantage in contributing to a culture of greater kindness. If you are the one to make an awful mistake, you might have a chance to redeem yourself by explaining that the harm you caused was truly not what you wished for.*

Certainly, what Bloom describes is part of it. Considering the intention of a person's controversial speech absolutely does amount to showing that person kindness, *and* you may be glad you did when a social justice sleuth finds your AOL Instant Messenger away messages from middle school and exposes the fact that you described your geometry homework as "soooo gay ugh lmao."

But honestly, it's so much more than all of that. It's about more than showing kindness to the could-be canceled, or even about creating a culture that could protect you from your own cancellation. It's also about creating a culture where fear of cancellation over unintentionally offensive speech stops paralyzing communication, making people afraid to have open, honest conversations or to make serious situations seem less scary by making jokes about them.

It's fine to get upset and even angry—but for the sake of everyone, you should keep in mind that the way you react could be hindering the opportunity for another joke or conversation that you might like or love, or even need.

A study conducted by James L. Gibson of Washington University in St. Louis and Joseph L. Sutherland from Columbia University and Princeton University found that 40 percent of Americans reported censoring themselves in 2019, with "worrying that expressing unpopular views will isolate and alienate people from their friends, family, and neighbors" being what "[s]eems to drive" it. As the researchers noted, this number is *more than triple* the 13 percent who reported self-censorship during the 1950s—the era of McCarthyism—which makes this 40 percent number even more startling.

It doesn't seem to have gotten any better since then, either. A 2022 study conducted by public opinion think tank Populace found that self-censorship was so "pervasive," "every subgroup had multiple issues with at least a double-digit gap between public and private opinion"—which creates a "false consensus in the public narrative" that can "drive false polarization, erode trust, and hold back social progress."

Thankfully, it seems I am far from the only one concerned about all of this. A 2022 survey conducted by pro–free speech group the Foundation for Individual Rights and Expression found that nearly six in ten Americans "feel that our nation's democracy is threatened because people are afraid to voice their opinions," and a 2021 poll by Harvard's Center for American Political Studies found that 64 percent of Americans reported considering a "growing cancel culture" to be a threat to their freedom.

A widespread fear of being able to speak has implications far greater than the problems it creates for professional comedians, because what is left unspoken can also be a missed opportunity for connection. Again: Whenever I've gone through something traumatic, it's always been helpful to remind myself that going through something awful, no matter how bad it sucks, also automatically connects you with everyone else who has been through it, too.

But what good is any of that, really, if we can't talk about those things at all out of fear that the Lindy Wests of the world will eviscerate us if we do so in a way that offends people, even accidentally?

Everyone has a unique perspective. If we don't let people share theirs, we are missing a huge opportunity not only for laughter, but also for learning. When we give people the space to take risks and speak freely, though, there are no limits for connecting with those around us—or for what we can find a way to laugh in the face of, no matter how otherwise distressing it may be.

CHAPTER 3

# DON'T ERASE ANYTHING

Being a woman is weird. As a younger woman, you constantly have to worry that every guy is just trying to hook up with you; as an older woman, you constantly have to worry that no guy is ever going to want to hook up with you again.

Unlike men, we don't have the option of knocking up a twenty-three-year-old at seventy. While men get to go from being "young and full of promise" to "distinguished and full of accolades," women have to go from "too young to be taken seriously" to "too old to be taken anywhere." Honestly, I sometimes can't believe that my husband, Cameron, actually married me when I am so clearly way too old for him. He's only two years older than I am numerically speaking, which means that, practically speaking, he's actually anywhere from ten to fifteen years younger.

Women have it the hardest of anyone when it comes to aging in our culture—except maybe for jokes.

Like, as a woman, at least I don't have to worry too much about being scrubbed from the face of the earth just because people have decided I'm outdated. I mean, I guess there's always a chance I'll get murdered, but most likely, I'll be free to keep living life until I die of something more peaceful. Like a long battle with cancer.

Often, jokes don't have that luxury, especially in recent years.

In June 2020, streaming services pulled episodes of shows (and, in a few cases, entire series) in the wake of George Floyd's murder because of scenes where white actors wore blackface or brownface. It was so widespread, in fact, that there was a 2020 article in Vulture titled "Every Blackface Episode and Scene That's Been Pulled from Streaming So Far." (I paid a fifty-dollar annual fee for a *New York* magazine subscription to be able to access it and include it in this book; you're welcome.)

*It's Always Sunny in Philadelphia*, for example, had five episodes pulled from Hulu that showed characters using blackface, brownface, and yellowface. Several of these featured an alcoholic, sociopathic, narcissistic, delusional character named "Dee Reynolds" attempting to showcase what she thinks are acting chops by performing characters she's developed—including "Martina Martinez," whom she describes in one episodes as being "a streetwise Puerto Rican girl who's always quick with a sassy comeback," and Taiwan Tammy.

Both of these impressions are *extremely* racist, using hack accents, wigs, and prosthetics. Dee uses brownface for her Martinez impression and yellowface (think big, fake buck teeth) for Taiwan Tammy. The over-the-top, blatant racism of the impressions, though, was clearly intended to demonstrate that such a lazy excuse for comedy was totally unfunny and pathetic. For one thing, Dee is quite obviously not supposed to be a role model character. She's set her college roommate on fire, kidnapped people, and destroyed the life of a priest to win a bet. What's more, her impressions do *not* go over well. At one point, for example, the glue-sniffing, illiterate, cat-food-eating alcoholic janitor Charlie Kelly's response to her performance is "This is so racist," making the point that even a glue-sniffing, illiterate, cat-food-eating alcoholic janitor realizes that lazy, hack racist caricatures do not equal comedy.

What changed between when they aired and now? Not the intention: deriding racism. Not the way society feels about the target: We still hate racists. What changed is that we think none of that matters.

That same summer, NBCUniversal completely erased (and I do mean "completely"—from syndication, streaming, *and* digital rental) four episodes of *30 Rock* that used blackface, at the request of creator Tina Fey and executive producer Robert Carlock.

In Fey's statement to companies that streamed or sold the episodes, she preemptively pushed back against any possible intention-based justification:

> *As we strive to do the work and do better in regards to race in America, we believe that these episodes featuring actors in race-changing makeup are best taken out of circulation. I understand now that "intent" is not a free pass for white people to use these images. I apologize for pain they have caused. Going forward, no comedy-loving kid needs to stumble on these tropes and be stung by their ugliness.*

(Note: In the previous chapter, I do argue for heavily weighing intent when responding to something that offends you, but that is different from calling for intent to be a "free pass." After all, I don't think that any decent person treats intent as a "free pass" in a situation where something goes wrong and someone gets hurt. Like, if I accidentally bump into someone, I will still apologize to the bumpee. Even if it's an accident. Even if it's an elevator.)

In any case, I don't think that intent is the most important thing to discuss when it comes to erasures—and that's not just because I talked about it the previous chapter, which means you've certainly already read it, because you're reading my book in order, probably in a single sitting, even if you were supposed to be at work hours ago, and even if your kids are crying that they're starving, because this book is just so captivating that you literally cannot put it down.

No. The most important thing to discuss here is that blackface is always good, and we should actually be adding *more* blackface into *more* episodes of television, certainly not deleting the precious blackface episodes that we have already!

That was, of course, obvious sarcasm *and* something that's probably already been picked up by Raw Story, which somehow took it seriously and published it with a headline about how I am a white supremacist piece of shit.

Because *of course not*, right? Of *course* I get that blackface and brownface are offensive, and also that they're offensive in a way that I will never fully grasp as a white woman. Personally, I've never painted my face black or brown myself, maybe with the exception of my run as Mrs. Dormouse in the Richmond Community Theater's production of *Alice in Wonderland* circa 1996, and that other time when I was like seven and wanted to be a black Labrador retriever for Halloween. (There are some uncomfortable photos of me eating candy after I had already removed the dog-ears part of my costume sitting in a family photo box somewhere. My brother found them recently, joking that if I crossed him he could publish them and have me canceled. I can see the Raw Story headline now: "A Fox News Blond Lady Has Been Racist Since the Second Grade. See the Shocking Photos Here!")

All stupid asides aside, it's just that, to me, all of that is irrelevant when we're talking about wiping things from history—because, regardless of its offensiveness or even its intent, erasing something limits our opportunities to learn. If the intent was positive, it's essential we see that for ourselves. If the targeting was wrong from the start, it's essential that we see that, too.

People who claim that our country has a racist and sexist history are, after all, absolutely correct. The thing is, though, if you want to work toward a greater understanding of these things, the last thing you should want is to erase the very examples that demonstrate it.

Personally, this is exactly why I wrote for *National Review* in defense of keeping classic fairy tales on the shelves. In a column responding to a library in Barcelona removing 30 percent of the books for children under six from its shelves over sexism concerns, I admitted that many fairy tales *do* have sexist messaging, but rejected that as being an argument for their removal:

> *I myself had some issues with classic fairytales as a child—exactly because of the way that they too often painted female characters as being fully completed by the companionship of a male character. My parents always tell me that, after the first time I finished watching* Beauty and the Beast, *I asked my dad: "Okay . . . so what does she want to do now?" Even as a toddler, I wasn't happy with that ending, because I couldn't imagine myself being truly fulfilled and "happily ever after" without having some kind of ambition to do something myself. (My parents should have guessed I'd grow up to be a thirty-year-old woman with multiple jobs who is not even close to getting married, but I digress.)*
>
> *Still, I don't think that these issues mean there is no value in these stories at all. They are enduring classics for a reason: They represent something about our culture and its past, and honestly, it is really never too soon to start having conversations about those things with your children. After all, that history does exist, and does continue to affect the present, whether you shield your kids from that fact or not. To me it seems clearly better to have those discussions rather than to hide your children from reality.*

It's the same with jokes. A joke, after all, can teach us a lot about what life was like during a period in time, even *and especially* if things have changed since.

Honestly, there are jokes that Past Me has made that Now Me considers hurtful.

Now, if you are a Media Matters for America researcher looking forward to this being the point in the book where I finally and truly admit that I made hundreds of thousands of dollars doing pro-Nazi comedy on plantations run by my slave-owning ancestors, I regret to inform you not only that my ancestors were Polish immigrants, but also that the example I will be using will be in regards to a much less touchy subject—death.

In the months right after my mom died of a rare illness when she was just fifty-seven years old, I would sometimes get a bit annoyed at people expressing their devastation about their dead pets on Facebook. My sister felt the same way as I did, and we would sometimes commiserate about our shared annoyance over the phone. I remember mocking someone in particular who wrote like six paragraphs detailing her anguish over the loss of her hamster. "*Wow, it must be so hard for her to deal with the fact that her hamster will never get to meet her future husband or go shopping with her for her wedding dress.*"

I eventually turned this into a bit I'd tell onstage:

> *I have a dead mom. And, right after she died, I really felt like I had it the worst, but then I was on Facebook and I saw someone had posted this picture of a bouquet of flowers with the caption "Rest in Peace Sara . . . 2000–2015. Such a tragedy." And I said to myself: "Oh my God, only fifteen? That's so sad! I've gone through* nothing."
>
> *And then, I keep clicking through the photos, and I realize that Sara . . . is a fucking* **dog**. *Like, a big dog too, like a German*

*shepherd. A German shepherd dying at fifteen is not a tragedy. That's fucking remarkable.*

That was how I felt at the time. But things have changed. Cheens—the same little kitty I told you all about earlier—has been with me through everything since I was a twenty-one-year-old Boston Market cashier, and he's older now. He has already been diagnosed with stage 2 kidney disease, inflammatory bowel disease, allergic bronchitis/asthma, herpes (it's a respiratory illness in cats; I was confused at first, too), and hypertrophic cardiomyopathy, which is a thickened heart muscle that can lead to heart failure or even sudden death.

Cheens's death is going to be brutal for me. I keep telling Cam that we had better save up some money because, as soon as he dies, I'm going to have to go to Passages Malibu after an inevitable mental breakdown. (Yes, I am also dealing with *this* by making jokes; it's what I do.)

The only consistent thing in the last twelve-plus years of my often-tumultuous life, after all, has been him sleeping underneath my chin. His purring has always been there to soothe me to sleep at night, and I can't even imagine what it will be like to have to live without him. I made a joke on *Gutfeld!* about how I won't let him die, no matter how badly he wants to—adding that I would put him on a ventilator or get him an ostomy if that's what it took. For years I've had to give him two medications via oral syringe every single day, and I will have to do so for the rest of his life. As of now, I also have to give him an asthma inhaler twice a day. Giving an asthma inhaler to a feral cat, by the way, is absolutely as hard as it sounds. Until I recently and finally got the hang of it, it was a two-person job that required Cam to keep him wrapped in a towel—a sort of kitty-burrito/makeshift straitjacket that keeps him from clawing his way out or wiggling away—while I hold the mask to his face with one hand and administer the medication with the other. But I will gladly do it, and anything else, because that is how much he means to me.

I would never describe Cheens as a "nice" cat. When we met, he was only six weeks old. Kittens aren't supposed to be separated from their litters that young, so he was never socialized properly, and he's bitten more than a few of my guests throughout the years. But he loves me, and he pretty much truly loves *only* me, so we have this really nice codependent relationship going on. I'm never going to have that bond with another pet. Sure, I have the French bulldog, Carl, that I got with Cam—he's adorable, loves everyone, and everyone loves him back. There's actually a very interesting ecosystem in the apartment because Carl knows how cute he is based on all of the attention he gets, and he just doesn't understand why I have this very obvious preference for mean, old Cheens. It's kind of like the two of them represent these two phases of my life—the feral street cat that I got when I was a broke cashier, and the fancy designer dog that I got with my husband. My life is far better now than it was then, both career- and relationship-wise, but still: Carl is *our* dog; all future pets will be *our* pets; I'll never have a companion that's just *mine* ever again. I'm Cheens's entire world, and our bond is irreplaceable.

In other words? The joke I used to tell mocking and minimizing the extreme levels of grief that people experience when their pets die has become extremely outdated to me personally. But that doesn't mean it was useless. It was funny at that time in my life, and it might be funny for someone else at that period in their life. If anything, knowing that I felt that way gives me information on approaching life going forward. If anything, it will probably make me think about how my eventual Dead Cheens Post (if I can even bring myself to make one before I am whisked away to Passages Malibu) could potentially impact people who could be seeing it from the eyes through which I saw similar posts when I made the joke.

We should reject calls for erasing jokes, even ones about far more controversial subjects than pets—yes, including extremely controversial ones such as race, and even in extremely controversial times like the summer of 2020.

For one thing, deleting old comedy over present outrages is really more of a cop-out than it is compassionate. Also in the summer of 2020 (the *New York Times* called it "the take-down summer for television" for a reason), Netflix deleted a 2015 episode of David Cross and Bob Odenkirk's sketch show, *W/ Bob & David*, over the use of blackface. The episode features a short sketch in which Cross plays a character named Gilvin Daughtry, who keeps driving up to a police checkpoint hoping to get evidence of police harassment on video. No matter how many times he drives up, the police just ignore him—until he drives up wearing blackface, at which point one of the officers asks the other, "Is this the guy who's been dicking you around?" The blackface-wearing Daughtry is then pulled out of his car, and a white cop tases him and sprays him with Mace.

Unlike the teams behind *Always Sunny* and *30 Rock*, Cross spoke out against Netflix's decision to remove the episode on *The Last Laugh* podcast, saying he was "surprised" and "disappointed," adding, "I know everybody involved with it was."

He continued:

> *Netflix isn't in the "we want to be friends with Bob and David business." They're in the international business of not upsetting people and continuing to get subscribers. So it's an easy, albeit thoughtless, decision to make. They're not interested in having a discussion about it. They've got a business to run. And so we were just the unfortunate recipients of that purge of anything that remotely, even in a tertiary way, approaches that idea. So it's all*

> *gone, it's just gone. They scrubbed the whole episode. They didn't even just take that sketch out. I mean, there are like three or four layers to the humor in it. And it's not somebody who's being offensive or doing it to be offensive, nor is the sketch itself trying to be offensive. And it actually makes a positive statement about the Black cop. I mean, it's just thoughtless and they just said, "I'm not interested in discussing what it means, get rid of it."*

He's right. In most cases, deletions are more about a business's bottom line than anything altruistic. If they cared about touchy subjects, they would *encourage* people to look harder at complicated issues. "Discussing what it means" is certainly a better option. Although Cross clearly stands behind his work, discussion being the better option is true even if you *do* find the material in question to be indefensible. I'm hardly the only person who thinks this way, either. In July 2020, television writer and cultural critic Alanna Bennett told *Variety* as much.

"It kind of freaks me out that [streaming services] can just pull episodes that have issues in them," says Bennett, "because I would rather the people of the future have access to the fact that the 'How I Met Your Mother' creators thought that a yellowface episode was okay in the 2010's," she says. "They had Jason Segel in full yellowface! I would just rather we have access to that history."

Several others quoted in the article expressed the exact same sentiment as she did.

Which makes sense! When we delete something, it robs us of the chance to see the controversy for ourselves, and to be able to have a well-informed discussion or debate about what happened and what it means. Ironically, Bennett's comments on that *How I Met Your Mother* episode provide a perfect example of this problem, because her recollection of it is actually incorrect. In the episode, Segel's character, Marshall, is actually not in yellowface, or anything close to it, himself. Rather, Marshall imagines that three of his friends are kung fu masters, and in his vision, those *other* characters are wearing silk robes and speaking in Asian accents—with one of them, Ted, wearing a Fu Manchu mustache. (The show's co-creator, Carter Bays, said the intention was "to make a silly and unabashedly immature homage to kung fu movies, a genre we've always loved," but ultimately apologized for offending people and called it a learning experience.)

The reaction at the time was mixed: Some people defended the episode, while others found it offensive and racist. As for any reaction to it today? Well, all of *those* opinions will likely be based on either a recollection of the episode from having seen it years ago (which might be wrong, like Bennett's was) or someone else's retelling of it (I got my information from a description and series of screenshots from a Screen Rant article). Obviously, neither of those options allows for as informed a discussion as the ability to view the entire episode would provide.

According to experts, losing some of these episodes forever becomes a real concern when streaming services remove them. Dan Wingate, a filmmaker and former technical specialist in film and television preservation at Sony Pictures Entertainment, told the *New York Times* that every time an episode is removed, there's an increased chance that it will be accidentally lost or destroyed, and then completely gone forever, especially considering the fact that many new employees in digital media might not have a firm grasp on the relevant technology.

"With new distribution companies popping up that don't have these asset protection structures in place, and metadata requirements for their inventories, a lot more digital assets could be lost," he said.

The only right, honest thing to do is to just keep everything. Erasing a joke to make the past look better amounts to lying about the past. It already happened, and pretending it didn't exist is no different than pretending that you *didn't* bang that guy who was popular in your high school in his parents' basement where he still lived after you saw him at your reunion.

The limitation of lying is that the truth is always going to be true anyway. The kinds of jokes that people told during a time can really tell us a lot about that time—and why wouldn't we want to know? As good as denial might feel, it's the same kind of Feeling Good as using drugs to escape the horrible reality of your life. It might make you feel better, but it doesn't *make* anything better, and actually keeps you from acknowledging the reality of the bad . . . something that's necessary for growth.

Although some people may see erasing what's become unacceptable as a sign of progress, signs of progress are exactly what you *are* erasing. Regardless of how you feel about a joke, regardless of how cringeworthy or even vile it may be, it is always better to have the option to confront and discuss it. Knowledge and information aren't the enemy—unless your aim is to delude.

CHAPTER 4

# NO ONE WANTS TO HEAR YOU WHINE (UNLESS IT'S FUNNY)

If you were alive during the spring of 2020, your spring was probably pretty rough. (If you weren't alive? Then you're either a baby genius or a spirit from the underworld. In either case, please reach out to me. I have questions!)

Even if you didn't have any of the Real Problems associated with the COVID pandemic—like someone close to you dying, losing your job or your business, or having to spend all of your time at home with your children—it still probably really, really sucked.

All of life's simple joys were suddenly either illegal or impossible, especially in cities like New York.

Want to go out to dinner? Too bad, that's illegal. Want to go outside and feel some fresh air on your face? Some dickhead is going to scream at you to put your mask on. It was terrible, and I can't even imagine how Club Girls felt! I mean, where were they supposed to wear their bandage dresses and heels *now*?

You guys remember places? We would go to them? Wild, I remember tweeting at the end of March. And the end of March was only the beginning!

Even though I was one of the Lucky Ones—healthy, employed, and childless—I was still going crazy spending so much time cooped up in the same handful of rooms with so few options to spike my own dopamine. It seemed like the only people in New York having any fun were the ones I saw injecting drugs with their friends during those Walks for My Mental Health that my therapist told me to take, and I can admit having flashes of desire to join them just to feel something other than a couch cushion under my ass.

Again: I *knew* that I was lucky, and I was intellectually grateful for that. Still, I found that a *rational* understanding of my comparative fortune wasn't always enough to quell my *emotional* distress.

The whole time, I kept seeing memes about how quarantine Wasn't So Bad: "Anne Frank had to hide in an attic, you just have to stay home and watch Netflix!" "Our ancestors stormed Normandy, you just have to sit on the couch!"

Like . . . okay? I am, in fact, *not* being hunted by Nazis. You got me there! But does that mean that I'm not allowed to be sad?

It truly was absurd. I saw other people make comments about how the pioneers had made it, so certainly, all of us would make it, too. As if that were an even moderately relatable comparison? Sure, I get it, the pioneers' problems were worse, but they were also literally the *opposite* problems. Pioneers probably *didn't* have time to notice the little annoying things about their pioneer family, because they had that exhausting, dangerous cross-country journey to distract them. Plus, if you were a chick, you were just constantly getting pregnant and giving birth, and then some of the kids you had already given birth to would sometimes die along the way before you reached your horse-driven destination.

So, no, they probably were *not* noticing how much their husbands' whistling bothered them because they were too busy discussing whether or not they should, I don't know, eat the body of the kid that just died in order to have the strength to pop out the next one.

At certain points, I got so empty inside that I remember thinking that I almost wished that I *could* fill the void with a pioneer baby—that the thought of a looming premodern medicine childbirth in the back of a buggy might have actually been good for me.

At the time, I wrote a column for *National Review* mocking the memes—which made me think about how big a role this exact kind of pressure plays in our culture. For some reason, it's near-universally accepted that the key to happiness is *positivity* and *gratitude* and "*If you woke up this morning with a roof over your head, then dance with joy that you are blessed enough to have a roof . . . even if your husband just moved out from under it to live with the secretary he has been banging behind your back for years because she is younger and better looking, and the kids keep asking. 'Where's Daddy?' Prayer hands emoji, sparkly stars emoji, prayer hands emoji.*"

Everything from social media to the posters in school hallways to the cross-stitched pillows adorning the couches of basic bitches seems to be riddled with quotations extolling variations on this principle, and they come from experts, thinkers, and celebrities ranging from new to ancient:

> *Be thankful for what you have; you'll end up having more. If you concentrate on what you don't have, you will never, ever have enough.*
>
> —Oprah Winfrey

> *Gratitude is riches. Complaint is poverty.*
>
> —Doris Day

*Wear gratitude like a cloak, and it will feed every corner of your life.*

—Rumi

*Reflect upon your present blessings—of which every man has many—not on your past misfortunes, of which all men have some.*

—Charles Dickens

*Appreciation is a wonderful thing. It makes what is excellent in others belong to us as well.*

—Voltaire

*Keep your face always toward the sunshine—and shadows will fall behind you.*

—Walt Whitman

There are thousands more, and they're more than just captions on the bikini photos that Instagram influencers post from their trips to luxury resorts. (Although they are, you know, *definitely* also that.)

For example, it's at the heart of movements like body positivity, or "Everything About Your Face, Body, and Hair Is Beautiful and Perfect," which teaches that your heart should be swelling with joy at the sight of your cellulite, stretch marks, and acne scars. Don't be embarrassed, these movements advise, about your fat belly . . . instead, just look at that belly and *love* it! *When you can look at your gut and say, Damn, that is* hot, *then you will be truly happy!*

On the surface, all of it makes perfect sense. I mean, hey—I definitely don't dispute the fact that the ability to view everything in a positive light would work wonders for a person's mental health. I don't doubt that being able to focus all of your mental energy on being grateful for what you have would transform your brain into a ray of sunshine that radiates inside and out. I don't doubt any of that, but my question is this:

*What if you can't?*

Gratitude and positive thinking can be helpful, but the pressure to practice them can also become toxic. It's not that I'm claiming to be smarter than Rumi. I mean, that guy has managed to be famous on Instagram without the aid of a single lip injection, which is impressive! And it's not that gratitude is bad. Although being thankful has not brought me the level of wealth that it has apparently brought Oprah, I still do make it a point to practice it as much as I can.

But at the risk of not making it onto the Oprah's Book Club list, I just don't think that gratitude and positivity should be heralded as *the* way to happiness the way that it often is.

For me, the reason is simple: It just doesn't work. Like most people, I have dealt with a lot of insecurity. For example: I'm extremely annoying, certainly far too annoying for someone with a chest as flat as mine. I was diagnosed with alopecia after losing tons of hair in my twenties—and, although it's gotten significantly better with the help of things like platelet-rich plasma (PRP), Rogaine, and no longer dating men who lifted my cortisol levels instead of my spirits, I'm always going to hate how thin it is without extensions.

Body-positive icon Ashley Graham (the Rumi of my generation?) suggests that the key is for me to simply decide to love the things I hate about myself.

"Your words have so much power," she says. "Every day, if you tell yourself, 'I love you,' if you give yourself one word of validation, it will change your mind."

(Note: Ashley Graham is a model. A thicker-than-most-models model, but a model nonetheless. She's literally hot *for a living*, so it's probably easier for her to convince herself that she is the thing she's getting paid to be than those of us who are still waiting for our first *Sports Illustrated Swimsuit* covers.)

Anyway, the truth is, I *don't* love everything about me, and I don't think I ever will. Does that mean I am holding myself back, squandering all hope for my happiness by staying stuck in a rut of self-flagellation?

Nah.

For me, it's been easier to take a more realistic (read: *actually possible*) approach—which is to not take myself too seriously. Rather than aiming to love the things that I hate, I've learned to love making myself and others laugh by joking about them.

When I wake up looking hideous, instead of exalting my reflection, I'll just snap a photo and send it to a friend with a joke about how I look like the mug shot of a woman who has just drowned her children. Then we'll both laugh, and I'll move about my day—not having *removed* the self-critical thought, but having *used* it as an opportunity for some laughter-induced, mood-boosting oxytocin.

Even comments about my insecurities hurt far less because I've already found ways to laugh about them. You want to tell me that I look like a little boy or tease me about wearing hair extensions? Tough luck. I've already made jokes on TV about how I look like a young Macaulay Culkin with a wig on. You want to inform me that the chest I see every day is small? Too bad, I've already joked with my friends that my bra size is "Double Mastectomy." I've already riffed onstage about how people with bodies like mine can't relate to the baseball analogy for sex—because everyone in my life has always just skipped second and gone straight for third—and I've already responded to a question from Greg Gutfeld about whether I'd ever been part of a hoax with "No. Wearing a padded bra, but other than that, no." During those despondent quarantine days, I may have been unable to achieve perfect positivity—but I was able to laugh by comparing my Unwashed Depression Messy Bun™ to one of those oil-soaked baby ducks on the Dawn "Wildlife Campaign" commercials.

I guess it could be nice to always see the good in a situation, but honestly? It can be even better to see the funny.

Plus, pressuring myself to focus on the good is just me setting myself up for failure—and there's a decent amount of evidence suggesting that I'm not alone.

Despite the popularity of body positivity as an *idea* over the last several decades, it doesn't seem to work *practically*. For example, a 2011 *Glamour* magazine survey found that 97 percent of women reported having at least one negative thought about their bodies every single day. A 2016 study published in the journal *Body Image* reported that only 26 percent of women and 28 percent of men describe themselves as being "extremely satisfied" with their physical appearance.

Loving every inch of your body just isn't something that a majority of people find themselves *capable* of doing, even though it's what we've all heard we *should* do. It's time to stop hammering the importance of something that's clearly impossible—and making already-bummed-out people feel the added burden of defeat when they fail to achieve it.

What's more, this applies to issues far beyond body image. For example, I personally have struggled with a cute lil' cornucopia of mental health issues throughout the years. I've been formally diagnosed (not to brag!) with ADD/ADHD, anxiety, and depression—and all of them have, at times, succeeded in making even the most basic parts of life feel impossibly difficult.

I can recall countless times when I've been deep in the depths of despair, feeling so goddamn awful and miserable that I'm unable to comprehend the onus of having to continue to live. Then, someone will tell me "Just look on the bright side:)" or "But you have so much to be grateful for!" and *boom* . . . my entire mood totally changes!

It changes, of course, in the sense that I *had* been depressed, but now I'm *also* experiencing extreme rage at the fact that you apparently think I'm so unconscionably stupid that I hadn't thought of that. That I wouldn't have decided to *just be happier* if that were an option for me!

Even if you're not stricken with anything clinical (sick brag, by the way!), I don't think it's possible for even the best-adjusted person to be positive all the time—and there is nothing that will frustrate you more than feeling like you have to censor your own feelings from yourself. Sometimes you won't feel optimistic. Does that make you an ungrateful, miserable jerk?

No, absolutely not. Especially if you can find ways to joke about the stuff that you hate about yourself.

In 2017, researchers from the University of Granada found "that a greater tendency to employ self-defeating humour is indicative of high scores in psychological wellbeing dimensions such as happiness and, to a lesser extent, sociability."

What's more, a study published in the *Leadership & Organization Development Journal* in 2014 found that a leader who used self-deprecating humor came off as more trustworthy and capable than those who told other kinds of jokes—or who used no humor at all—even though "there were no differences between the conditions in ratings for how funny the leader was."

Sure, there are those who don't buy it. Apparently, especially Gen Z. A quick Internet search on the topic reveals that there are a number of college students who felt the need to write columns about how self-deprecating humor isn't funny or helpful at all because there's nothing nice or cool about being mean to yourself ever, mmmkay?

But that just isn't true. Some things suck, and sometimes those things are yours. If you do have to live with them, why not at least disarm them? They may not be ideal, but why not make them less big *of a* deal by joking and laughing about it all?

Throughout her life, Joan Rivers seemed to have mastered this sort of humor. Although Rivers certainly did her fair share of mocking the appearances of other people ("You're sending a message out to people saying it's okay, stay fat, get diabetes, everybody die. Lose your fingers," Rivers said to Howard Stern about Lena Dunham's constant nudity on Dunham's show *Girls*) but perhaps her favorite target for these jokes was herself:

*I've had so much plastic surgery, when I die, they will donate my body to Tupperware.*

*My breasts are so low, now I can have a mammogram and a pedicure at the same time.*

*I was so flat, I used to put Xs on my chest and write, "You are here." I wore angora sweaters just so the guys would have something to pet.*

At times, these jokes could get really dark, such as:

*My husband killed himself. And it was my fault. We were making love, and I took the bag off my head.*

In fact, Rivers constantly credited her sense of humor as the way she got through the toughest parts of life.

"I purposely go into areas that people are still very sensitive about and smarting about, because if you can laugh at it, you can deal with it," she said. "That's how I've lived my whole life."

"Love your life and yourself, and you will be happy!" may make sense intellectually, but in practice, it isn't so easy. That creates a lot of pressure for those of us who can't seem to do it, which is likely a lot of us. After all, many of the same people who preach positivity with their words blatantly counter it with their actions.

This hypocrisy was something Rivers pointed out in regard to Jennifer Lawrence, saying, "I love that she's telling everyone how wrong it is to worry about retouching and body image, and meanwhile, she has been touched up more than a choirboy at the Vatican."

I'm all for empowerment—to me, though, there's nothing more empowering than learning that you can absolutely hate things about yourself without letting that hatred consume you. It's actually realistic, and so freeing to realize.

The even better news? Your jokes really don't even need to be that funny in order to be helpful.

For example: Because of my ADD/ADHD and anxiety, I found that I spent too much time in the quarantine stuck in a terrible cycle: getting distracted and dicking around doing things that weren't work, freaking out about the wasted time, and then compulsively trying to escape my anxiety with the disgusting, compulsive habit of squeezing any clogged pore on my face or body that I could find—wasting even more time, while also making my skin look like I'd been blowing clouds behind a dumpster all night. One day my husband started referring to my activities as "dickin' and pickin'." He would come out of his office and ask me, "What are you doing out here? Dickin' or pickin'?"

Was it funny? No, not really, and honestly, my husband pretty much never is. But I quickly adopted the phrase anyway—and for some reason, using such ridiculous, silly wording to describe such frustrating issues gave me some relief from the turmoil, making it easier for me to deal with.

Pushing for people to think and speak positively all of the time may sound like a good movement on the surface, but really, all it is is a form of censorship, scolding them for being honest about the way that they actually feel.

The adage may say that no one wants to hear you whine, but that actually isn't true. The truth is, no one wants to hear you whine *unless it's funny*—and, in order to feel better about something you feel bad about, the only person who really needs to find those jokes funny is you. And if another person with the same insecurity *does* find your joke funny? Well, not only will you have made them laugh, but you'll also have made them feel less alone.

CHAPTER 5

# SHITBAG

Brutally early on the morning of November 29, 2020, I had horrible, and I mean *excruciating*, stomach pains. The kind of "excruciating," in fact, that made me regret ever having used the word "excruciating" to describe other things in the past. This pain was unlike anything I had ever felt before.

It was difficult to walk; it was difficult to breathe, and I told my then-fiancé Cameron that I needed to go to the hospital *now*.

Thinking that I was probably just overreacting (a fair assumption, as anyone who has had the displeasure of living with me will tell you), Cam told me to just lie down and sleep it off. When you hear the rest of the story, you will understand why he wound up riddled with guilt over his suggestion, but I do understand where he was coming from. It was peak pandemic! At this point, you couldn't even go to a restaurant in New York City, and, pandemic aside, the last place anyone should ever want to go is a New York emergency room early on a Sunday morning. If you haven't been up on drugs for at least five days, you're just not going to fit in.

But we went. We hopped in an Uber to Mount Sinai West,

and he carried me inside to become yet another ailing cog in the Emergency Room Crazy Machine.

While I was there, the pain just got worse. I could just barely stand for the X-ray, but by the CT scan, I couldn't even manage to roll myself onto the table. (And that was *after* I was given a morphine drip.)

I could feel myself deteriorating *so* quickly, and I'd never felt anything like that before. I told Cam that I was dying, and I was certain that I was. He thought I was being insane, and just kept asking me what kind of food I wanted to order when I got home. The X-ray had turned out okay, and certainly the CT scan would show the same thing. I was a perfectly healthy thirty-two-year-old woman! Everything was going to be fine.

The bad news? I was right about the dying.

Healthy thirty-two-year-old or not, my CT scan revealed that I had a perforated bowel. I didn't have long to live without an emergency surgical procedure called an "ileostomy," which is, essentially, a fancy word for "shitbag."

I myself learned the word "ileostomy," by the way, as the surgeon was telling me that I needed one. He explained to me—in that calm, clinical way that a doctor can explain absolutely anything—that I would have to go into surgery to get a bag put on me to "empty my bowels into" while my colon healed. That way, none of the "fecal material" would leak into my bloodstream and make me septic.

Cameron was still not getting it. He actually asked the surgeon when we should come in for the operation! Given the fact that we had *just* been having a perfectly normal Saturday night, I can absolutely understand this. That's how life is: Sometimes things can change faster than you're able to keep up with accepting them. At the time, though—as I was withering in pain and fear—I remember looking at him exasperated.

The surgeon, still casual, cool, and clinical, just looked at him and said, "We are prepping the OR."

The surgeon told me that I'd likely have the bag for anywhere from three to six months. (For the first time, I was actually relieved that then-governor Andrew Cuomo had forced me to cancel my thirty-person December 6 wedding.)

They told me they wouldn't be waiting for the results of my rapid COVID test. Before I could even google the details of what was about to be done to me, I was told to hand my jewelry to Cam.

Because I *hadn't* had the chance to google, I didn't realize that I was, in that moment, even closer to death than I'd thought. Because I hadn't googled, I didn't realize that I actually wasn't going to be fitted with any sort of waste-disposal apparatus the way I'd assumed—just my actual small intestine hanging out of a gaping hole in my abdomen, randomly spurting out waste. (I didn't know this until a few days after surgery, actually. That's how long it took for me to have the strength to look at it.)

As you might guess, I was extremely nervous. What might surprise you, though, would be what was freaking me out the most.

*What am I going to tell people? I am going to be in the hospital for at least a few days, and recovering at home for at least a week. How am I going to explain something so disgusting and weird to anyone? What am I going to tell Greg? What am I going to tell my dad? What are people going to say? How am I going to explain this to anyone? What are they going to think? Of all the random traumas that could have happened to me today, why did mine have to be so niche?*

Thoughts about what I might say and what others might think—not thoughts about my actual life, health, or well-being—were the ones filling me with dread as I was wheeled into surgery. There's no good script, after all, for how to talk about the things no one wants to even think about.

I wound up being in the hospital for three nights and four days.

"Because of COVID," I wasn't allowed to have more than two visitors per day—for a maximum of two hours each—between the hours of 10 a.m. and 6 p.m. (I used those quotation marks because, as far as I'm aware, there's no evidence that COVID becomes more contagious or dangerous after the hours of 6 p.m. EST.)

It sucked. I mean really, really sucked. In theory, a building full of people on opiates wearing open-ass gowns could be a fun concept, but the hospital just does *not* do it well. Truly, it is a fucking awful hang. I was shocked, uncertain, and afraid, and I had to spend most of my time alone. Mostly, my only company was the series of roommates that I had on the other side of the curtain.

I'm not certain what my first roommate was in for, but whatever it was, I think its prognosis improved drastically the more that you passed gas. I don't know for sure. I'm just guessing this because she spent the entire evening loudly farting, and then singing hymn-style songs that she'd made up herself, giving glory to God for those farts. I tried to turn up the *Forensic Files* I was watching on my tiny TV to drown out the noise, but *she* told *me* to turn it down.

With the kind of out-of-fucks-to-give-ed-ness that can come with having your intestine hanging out of your body and an IV drip of Dilaudid in your arm, I asked her just how in the fuck she could possibly ask me to do that when she'd been fart-singing all night. She apologized and said that she didn't think I could hear her, and we wound up actually chatting until she was discharged in the morning. I still have her number in my phone.

The next woman, I wouldn't chat with. I wouldn't be able to, really, because she was in terrible, terrible shape. She was in her forties, completely alone, and absolutely ravaged with cancer. She was crying as they told her that, not only had they taken out her uterus and her ovaries, but they had also taken out pieces of her liver, stomach, and diaphragm, because they'd found cancer there as well. She asked if she was dying; they said that they didn't know yet. That Tuesday night, she had low blood pressure and a fast heartbeat, so doctors were rushing in and out all night long. I had an hour where I couldn't get pain meds because everyone was busy attending to her. I remember feeling miserable, and then guilty for feeling miserable, because it's not like I had things as bad as she did.

I completely understand why people in hospitals need advocates, lucid people by their sides for more than a handful of hours per day. There were times I was so, *so* dirty and too doped up to even realize it, until Cam came to visit, got upset, and made them change my gown and sheets. I couldn't bother to eat, so both he and Keith—a gay gymnastics coach who was living with us at the time, would later be the flower girl in our wedding, and will always be my best friend—would come bring me food and feed me.

Honestly, what got me through was trying to laugh or make other people laugh. When Cam, for example, brought me broccoli cheddar macaroni and cheese instead of the broccoli cheddar soup that I wanted from Panera, I loved the laugh I got from both him and the nurses as I told him, "It's fine, it's only the one thing I had to live for all day."

I couldn't wait to get out of there. In addition to the shitbag, I also had a drain inserted into each of my sides, to collect the fluid from my wounds. I really was just walking around wearing a hula skirt of my own bodily fluids for several days.

On day four, they finally told me that I could go home. They just had to take my drains out first, and would be right back with some medication for the pain of that removal process. When I asked them how long the medication process would take, and they said twenty minutes, I said:

*Just do it now.*

They did! I remember Cameron looking at me like, "Who is this chick?" as I had them ripped out, screaming "FUCK," completely and totally conscious and aware.

Anyway, despite earlier estimates, I wound up needing the bag for only five weeks. I say "only" like it wasn't one of the most tragic experiences of my life, but it was.

Sure, it got easier with time. But it was also really, really hard, especially at the beginning. In case you don't know, let me tell you that it is *strange* to sit there and see your shirt moving, to hear something underneath it making noises, and to know that that means you are essentially involuntarily shitting as you sit there. Especially if you're at work, and very especially if "work," for you, means being on television.

Plus, when I finally *did* have time to google it, everything I read made it far worse. Because all of it was like:

*There is* nothing *bad about having an ostomy! It's just the stigma of them that makes you feel bad! I climbed Mount Everest with mine! Here are the beautiful photos! My ostomy is the best thing in my life; it makes me so happy. Sending love to my fellow ostomates; ostomies are hot!*

And I'm just sitting there like . . . who wrote this? Did a *stoma* write this? Because, honestly, I had more fun at my mom's funeral than I had the first few weeks that I had that bag. And I didn't even kill her!

God, I'll never forget how bad some of those times were. I couldn't figure out how to make it stop leaking, and every time it did leak, I had to bother Cam to help me change the bag. Props to Cam for still marrying me after all the times he had to see me walk out of the shower with my intestine out, trying to catch all of the waste that was spewing from my stoma with my hands so it wouldn't get on the rug.

I had constant anxiety attacks, because the sight of my small intestine hanging out of my stomach freaked me out, but there was nothing I could do about it, because you can't run away and escape from your own body. I was having more and more trouble coming up with excuses for why I couldn't see most of my friends—because there just was really no chill way to admit to them that my colon had exploded, I had almost died, and I was currently living life as a poop-smoothie machine on the fritz. I couldn't bear the additional anxiety of what was guaranteed to be an arduous, humiliating conversation. I worried constantly that they were all mad at me or hated me.

One night, I had a leak for the second time in just a matter of hours while I had two of the few friends that I *had* managed to tell over to visit. I had to spend three hours just sitting there on the couch, marinating in my own itchy shit bandages (if reading this grosses you out, just imagine how grossed out I felt living it), waiting for the nurse to come over and try to figure out what was going wrong. Thankfully, those two friends were able to joke around with me about the whole thing after the nurse left. I remember feeling so grateful that they were treating me normally as I sat there, ate fried chicken, and talked about my plans to throw a small "Kat's Out of the Bag!" party after this was over—where my guests and I would bong champagne out of any unused-but-unpackaged bags. (Note: I absolutely *did* do that.) It may seem like nothing, but I'd been feeling like a walking, spewing science experiment, and the last thing I needed was to feel even weirder because people seemed afraid to talk to me the way they normally would.

A lot of those early days were really, really awful—so you can imagine how much worse I felt when I searched the Internet looking for some form of schadenfreude and found only stories of people who were managing to conquer monumental athletic feats with their bags, while I couldn't do anything but lie perfectly horizontally (the wrong movement might cause another leak!) and binge-watch Jodi Arias documentaries. (I mean, Travis Alexander's roommates' not noticing the smell of his decomposing body in their house for five days really does tell you everything you need to know about men.)

I'll never forget the day that Keith had to learn how to change the bag, because Cam had a meeting that he could not miss, and we had to have a nurse over to help because that leak had been the fourth one in about sixteen hours—including once while Keith had taken me to get my nails done to "get my mind off of things," and another one thirty minutes before I had to leave to film *The Greg Gutfeld Show*. We didn't want to put it on ourselves again, because apparently we were doing something wrong, as you're only supposed to change it every five to eight days.

Cam and I got into a fight about his work/life balance, with him telling me he had to do this meeting.

So, it wound up like this: Me, lying on top of a puppy pad on the bed, with both my vagina and shit-covered colon out in full view, and Keith standing there in all of his little-shorts, gay-gymnast glory, wearing elbow-length rubber gloves and trying to learn from this random nurse lady as I continued to weep and wail. (And Cam, of course, doing his best in that Zoom meeting just down the hall.)

If that scene sounds horrific, you're right. What *did* help was being honest about how much it sucked. If it also sounds kind of funny? Well, you're right about that, too. The whole thing was, in fact, *so* absurd that I *had* to laugh at it; I can't imagine how much worse it would have been if I weren't laughing.

I found a lot of ways to laugh during that time. For example, saying that at least I could now start an OnlyFans account and no one could ever call me slutty because I'd be *breaking the stigma* with my massive ostomy scar. (Yeah, I know, I could get it lasered off. I have certain friends who remind me of that all the time, but they are also the kind of people who would get *theirs* lasered off, only to spend money getting a tattoo to encapsulate what their *ostomy journey* meant to them.) What helped was Greg Gutfeld texting me: This is your Vietnam. What helped was naming the bag Beth, and joking that, for weeks, every time I had sex it was the world's worst threesome. Perhaps the funniest thing was the instructions relating to sex: "**DO NOT ATTEMPT TO PENETRATE STOMA!**"

I mean, the only reason that those would even be there is that some people had needed to see it. Worse: the only reason that they would need to be bold, underlined, and in all-caps would be that *many* people had needed to see it. (Actually, maybe *that* tells you everything you need to know about men.)

The very first time the bag exploded, it happened overnight. I woke up to discover it on December 6—the morning I was originally supposed to get married. Waking up to an exploded ostomy bag was definitely horrific, but I still couldn't help but laugh at the irony of it all. I'd had to trade in my wedding dress for a shitbag in the blink of an eye! I immediately texted the few close friends that I'd told about the bag:

> Dec. 6 Expectation: Marrying the love of my life in an intimate ceremony overlooking the Hudson River, surrounded by my closest family and friends.
>
> Dec. 6 Reality: Waking up to realize my shitbag had exploded.

By the way, one good thing about all of this was that it showed me I was definitely making the right call by marrying Cam. I mean, I didn't have doubts in the first place anyway, but most couples don't get the chance to get tested with something like this before they get married. With this situation specifically, most couples generally need to wait until around age eighty.

Probably the hardest I ever laughed, though, was during the road trip that Cam, Keith, Carl, Cheens, and I took from New York City to Detroit to see my family for Christmas. I had eaten as many Mike and Ikes as I could in an attempt to "stop up" the flow of the stoma, but I'd still have to stop along the way to empty it twice.

While I was emptying it in a Pennsylvania rest stop, the most exuberant rendition of "Walking in a Winter Wonderland" that

I have ever heard in my life (like, it sounded as if the original version had gotten into MDMA and body paints) started blasting over the radio.

*A beautiful sight! We're happy tonight*
*Walking in a winter wonderland!*

Those sprightly, singsongy lyrics, juxtaposed with my trying to empty the contents of the bag into the toilet without splashing them everywhere or stepping into trucker piss, made me start laughing *hysterically*—and I didn't stop for at least an hour. I still laugh at that sometimes, and I probably always will.

Now, if anyone reading this ever needs an ostomy, I want to be clear about something: It *did* get easier. After losing hope because countless nurses (and several doctors at the emergency room when I went back because I thought my surgery site was infected) had told me that there simply "wasn't enough surface area" on my body to reliably prevent leaks, I was actually able to get a new bag system at a follow-up appointment with my surgeon that made them far more rare. I was able to enjoy the Christmas holiday at home with my family—laughing, eating, playing games, and even doing some shopping—without a single leak. Although I don't think I'll ever have it in me to climb Mount Everest, I eventually started to almost get the hang of the whole thing. I mean, it's not that I don't understand what all of those Ostomy Positive Influencers were trying to do. They were, of course, trying to say that you can still live a full, happy life . . . even with an ostomy. Still, the fact that they kind of started there, without acknowledging having ever had any sort of real struggles, made me feel like there was something wrong with me for not even being able to climb onto the couch without risking a leak.

At the end of December 2020, I had a follow-up CT scan to

evaluate whether I was ready for reversal. I wasn't nervous, because I'd had CT scans before. That was wrong of me, by the way, because calling this horror-show procedure a "CT scan" is deceptive at best. Really, a better word might be "cruel." When I got there, they told me I wouldn't have to get an injection of contrast, and I was relieved for two seconds . . . until I learned that they were going to pump the contrast directly into my asshole instead. I can't describe the pain to anyone who hasn't experienced it, but just imagine the worst, near-death-food-poisoning-and-the-flu-at-once diarrhea that you have ever had—and you not only have to hold it, but you're also getting more and more of it pumped into your bowels.

I got the okay, and found out I'd get to have the surgery at the beginning of January. I was so, *so* excited. Probably how people feel when they're about to get boobs.

The reversal, unfortunately, was extremely difficult. I wound up having complications! Basically, there was an issue with one of my staples, causing me to gush massive amounts of blood. So much, in fact, that they wanted to measure it—which meant that I'd have go into what was basically a bucket with handles next to my bed, just a few feet away from a crazy old lady who was freaking out that the doctors were "experimenting" on her because they wanted to give her potassium supplements, and, even worse, called taking a shit "making poopie." (I know this because I had the pleasure of overhearing her "poopie" the bed multiple times.)

Eventually I lost so much blood that I wound up needing a transfusion. And guess which day it was that they told me this? January 6, 2021.

Yep.

Keith was there visiting me, and as we watched in horror at rioters storming the US Capitol, we couldn't help but laugh about how, no matter what, I'd always be able to answer the question "Where were you on January 6, 2021?" with "Shitting straight blood into a bucket in front of my friend Keith, with just a curtain separating me from a crazy old lady who was explaining to no one in particular that 'everyone liked Trump at first, but now they don't' and 'all those people who were storming the Capitol were *clearly* not Republicans or Democrats, but *Russians*,'" and I'd be *telling the truth.*

Whenever I see an article or news clip about AOC sharing how traumatic January 6th was for her, I sometimes can't help but send it to a select few friends along with the comment, "I bet I had a worse January 6 than she did."

I passed out during the transfusion, and woke up to find that someone else's blood had exploded all over me while I slept. It sounds horrific, and it was, but honestly? Maybe it was the pain meds, or maybe it was just all that I'd been through already at that point, but I started *laughing.*

I stayed home for two weeks after that. The wound was so huge at first that I couldn't even see my belly button. The "bandage changes" I had to do twice a day were, much like the CT scan, deceptively labeled. The bandages were packed deep inside the wound, and the "changes" involved Cam ripping those out, me soaping up the open wound, then him using a stick to shove saline-soaked new ones in there. We had to do this twice per day, and I still wonder what my dad, who had come to visit to help take care of me, thought about having to hear my screams.

The whole thing was pretty bad. But not putting any pressure on myself to pretend otherwise made it a lot easier, and so did being able to laugh at it. I mean, I know that the saying goes "Tragedy plus time equals comedy," but a lot of the jokes in this chapter are actually things that I wrote while I was still going through this stuff. It was a terrifying experience, but honesty and laughter really helped me to overcome my fears.

The more I think about it, the more insane it seems that the thing I had been the most concerned about on the day I almost died was how I would *tell people* about it. I mean really, how distorted is that?

Tragedies, especially weird ones, can be so isolating. For that reason, the friends I found myself wanting to spend the most time around during my ostomy experience were not the ones with the most conscientiously chosen words. I saw that as a sign that they were uncomfortable, and that made *me* feel uncomfortable in return. I mean, I felt like an absolute freak during those five weeks already—and the last thing I wanted as I was going through it was for anyone to reinforce those feelings by treating me any differently than they had when all of my innards were inside of me where they belonged.

Now that I think about it, I remember a moment I had with my mother a couple of days before she died. We were the only two people in her hospital room, and she turned to me and said, "Katherine, you know I'm dying, right?" I told her that, well, yeah . . . I did. She *laughed* and said, "Thank God. Everyone else is treating me like I'm either a kid or a retarded person." (I *do* worry that people might try to cancel her for her use of the word "retarded." All of you, please rest assured that cardiac amyloidosis did a pretty good job of canceling her already.)

Here's what she meant: Even when it was beyond obvious she was on her way out, everyone she talked to kept insisting that she was going to be better, that she was going to be just fine, that she was going to be eating kielbasa soup with us come the next Easter. Even my dad was doing it. A few nights before she died, I remember being so frustrated with hearing my father's delusional optimism that I started throwing cutlery around at the table. (I should be embarrassed, but it's a restaurant right outside a top hospital; I'm sure they see few people who are *not* living in an emotional meltdown.)

Rationally, we all know that miracles are called "miracles" for a reason. They defy the odds and even science; we all logically understand this. Still, for some reason, when we hear that people are sick, we treat them as if *they* do not know this—which, whether we like it or not, inherently means we are treating them differently just because they *are* sick.

Just think back to August 28, 2020. That day, the world was shocked to see a statement posted to Chadwick Boseman's Instagram announcing that the *Black Panther* star had died of stage 4 colon cancer. It was shocking, of course, because it was the first time any of us had even heard that he was sick.

But Boseman, the statement explained, was actually diagnosed with stage 3 colon cancer in 2016—the year before he had even started shooting *Black Panther.* Boseman had been battling this illness the whole time he was filming, and almost no one knew it. Really, almost no one—it wasn't just the fans who were left in the dark. *Black Panther* cowriter and director Ryan Coogler had no idea, and neither did *Da 5 Bloods* director Spike Lee.

One of the few who did, his longtime agent Michael Greene, explained Boseman's reasoning this way: "Chadwick did not want to have people fuss over him."

He wasn't the only one, of course. Actor and comedian Norm Macdonald did the same thing. On September 14, 2021, we found out that he had died of a cancer he'd had for nine years, but had chosen to hide the entire time—because, again, he didn't want people looking at him any differently.

It's not just famous people who struggle with this, either. As of this writing, a Google search for "How to tell people you have cancer" produces a staggering 965,000,000 search results, and "people treat me differently because I have cancer" produces 914,000,000. To put that in perspective, the word "cancer" itself only produces about 779,000,000.

All throughout the Internet, there are countless different people with the exact same concerns and complaints: Their illness is causing them extreme amounts of pain, sadness, and fear, and it totally sucks how, on top of that, everyone they know is treating them all weird because of it.

When someone is suffering from a grave illness, it is in many ways understandable that the people around him would feel compelled to speak to him in a way that matches that gravity. The truth is, though, that there is a lot of evidence suggesting that the opposite of gravity, humor—yes, *humor*—is actually incredibly helpful. One (albeit) small study of 340 terminally ill people found that 93 percent of them said that keeping their sense of humor was "very important . . ." ranking its importance as highly as the absence of pain. A 2018 paper titled "Humor Assessment and Interventions in Palliative Care: A Systematic Review" took a comprehensive look at thirteen different studies on the subject and found that "humor had a positive effect on patients, their relatives, and professional caregivers. Humor was widely perceived as appropriate and seen as beneficial to care in all studies."

In other words? Everyone says that a terminal illness is *not* a situation for jokes—except for most of the people with terminal illnesses, and the experts who study them.

It may seem counterintuitive, but to me, it makes sense. My own medical problems, of course, never reached the level that Boseman's or my mother's did. There's no comparison; that's obvious by the fact that I'm still aboveground. But the principle remains the same: Treating a sick person as if they suddenly live outside reality—or any differently than we'd treat them if they were perfectly healthy—gives their illness even more emphasis than it already has.

Think about it: When we make sure to carefully craft every interaction that we have with someone because they're sick, what we are essentially doing is sending the message that they *are* their illness. Or, at the very least, that their illness is what carries the most weight—more weight than the entire rest of their lives, regardless of how long they may have lived without it.

The truth is, people who are suffering with any kind of affliction are still people. What's more, it doesn't mean that they don't want to laugh. In fact, it might even mean that they need to.

CHAPTER 6

# LIVE, LAUGH, DIE

My mom's mom died in February 2015, just three months after my own mom did. All of it was pretty terrible: My mom dying only three weeks after she was (finally) diagnosed with cardiac amyloidosis, going to her funeral, and then having to go to my grandma's funeral at the same Hamtramck, Michigan, church just a few months after. Then I had to go to my grandma's wake in the same hall. My family, my then-boyfriend, and I had to sit at the same table as the same funeral guys who had embalmed them both. No wonder I got so drunk.

A few days after *that* emotional minefield, I was in the makeup room getting my hair done to be a guest on *Red Eye*, months before I worked at Fox News, or even knew that *The Greg Gutfeld Show* was something that would ever exist.

I had come, by the way, straight from a travel experience that honestly rivaled *Planes, Trains, and Automobiles*, minus the happy ending where I gain a lifelong friendship with John Candy. My Spirit Airlines flight had been canceled, and I didn't want to miss the chance to be on *Red Eye*, so there was only one option: The boyfriend and I would have to take a combination of buses and trains to make it in time for me to appear on the show. (When I say "bus," by the way, I *do* mean Greyhound bus, which is something I used to do all the time as a broke comedian. The worst one was probably the one I took alone that left out of Pittsburgh at 3 a.m., where I had been doing spots for The Exposure. It got into Union Station in Washington, DC, Monday morning just in time for me to change in the bathroom of the Potbelly Sandwich Shop and take the train directly to my job at the Leadership Institute's *Campus Reform,* in Arlington, Virginia, where I would go on to have a day that I would definitely not put on my résumé. If you've never been on a Greyhound, I'll just describe it like this: The entire time you're sitting there, you're saying to yourself: *I don't know where all of these people are going, but for their sake, I hope it's the hospital.* I'm not judging, either—I'm sure everyone else on there thought the exact same thing about me, and half the time, I'd probably have been better off if I had been.)

Anyway, the conversation with the hairstylist went like this: She saw my suitcases, asked me where I'd gone, and I told her I'd been home in Michigan. She said why, I said my grandma's funeral—and then I felt that all-too-familiar, sucking-the-air-out-of-the-room feeling as she told me, "*Oh my God, I'm so sorry.*"

Then it got worse. She asked if it was my mom's mom, or dad's mom, and when I told her it was my mom's mom, I prayed to a God that I don't believe in that she wouldn't follow it up with exactly what she said next.

*How does your mom feel about it?*

Fuuuuuuuuuuck.

I was coming off a rough trip, off a rough weekend, off a rough couple of months . . . and now I was faced with this. I had to either tell her that I actually had no idea how my mom felt about it—because she'd been in the ground for three months and had decomposed far beyond any capability of feelings—or I had to lie. I went with the former (although I still wasn't confident enough with expressing grief to put it *that* bluntly) and wound up having to feel far worse than I'd already been feeling all day. Now, in addition to feeling sad that the people around me kept dropping dead, in addition to dealing with the stupor and general sense of defilement that comes from a night of bus sleep, I also had to feel like I'd killed the vibe in the greenroom and had made everyone miserable just because my life was happening the way it was happening.

It would have been better if it could have been a normal conversation—if I could have made jokes about how I had probably spent more time eating pierogi in a Hamtramck basement with embalmers than any other twenty-six-year-old girl in the country, if the topic hadn't been one we are so obsessed with talking about "correctly" that everything we allow ourselves to say about it is wrong.

We do, to be clear, do that. If you don't believe me, the next time you're at a party, just try breaking a small-talk silence with the question "So, who here do you think is gonna die first? It's gonna be *someone*. Who do you think it's gonna be?"

The person you're talking to will get weird. Trust me; I would know. I just did it again last week.

But here's the thing: I did it not only because I am irreparably awkward, but also because I really wish that it could be acceptable to talk about dying and death more often, more honestly, and more casually than we do. Not just for my sake, but for everyone's. Our fear of talking about dying and death—our paralyzing ourselves out of our fear of saying something "wrong"—not only hurts the people experiencing grief by making them worry about making other people uncomfortable, but also hurts anyone who is ever *going to go through it*, because it does a piss-poor job of preparing people for what it actually looks like. That's especially unacceptable given the fact that anyone who lives even a quarter of the typical life span will probably go through losing someone they love—and, of course, the fact that we ourselves are all going to die. It's true: When it comes down to it, other than birth, the only thing we have in common is death.

Because of the way that everyone insists on whitewashing it, though, people don't know what death is really like until it smacks them in the face. I know because I had my own face-smacking the last night of my mother's life. Thanks to what I'd seen in movies and on television, I expected a lot more. I expected that I would be spending those final moments having important, meaningful conversations with her—ones where we would solve all of our issues and she'd leave me with some kind of beautiful, thoughtful, flowery words of wisdom and inspiration that I could carry with me for the rest of my life—in an ambiently lit room with soft piano music playing in the background.

It wasn't like that. First of all, the two lighting options in a hospital are "Off" and "Have I Always Been This Ugly?" and the closest thing to background music in an ICU is all of the beeping, sometimes communicating to nurses that one of the keeping-alive machines is losing. As for the conversation? My mom was a little too, well, *dying* for flowery pronouncements, so the majority of our final exchanges of words consisted of her telling me how tired she was, or asking me if she could have some ice chips because she was thirsty. I pushed so hard for the kind of deep talks that I thought we were supposed to be having, and it wasn't until years later that I realized how quixotic my expectations had been.

I couldn't help but think back at all of the crap fed to me throughout my life that had created those expectations. For example, *Titanic*. Like, there's no way that Jack and Rose would have had that meaningful, emotional exchange as he was hanging on to that lifeboat. Jack was, after all, dying of hypothermia—so he would have been incoherent and babbling if he could have even spoken at all. Then, I couldn't help but think about how I wasn't even sure if that's how the scene went, how my mom never let me watch *Titanic* like the rest of the girls in my Girl Scout troop in fourth grade, all because there were boobs in it. (I'd repeatedly argued with her, telling her that she should let me watch it because I was going to have boobs someday. Turns out? Joke's on me.)

I'll never forget the first Mother's Day without my mom. I escaped to Cape Cod with some friends, but unfortunately, Instagram followed me. I felt miserable. Scrolling through all of the tribute posts to living moms, seeing people gathering for brunch the way you only can with those loved ones who happen to still live aboveground.

The subsequent Mother's Day, though, I decided to do something different. I broke up everyone's feed by posting a picture of my laundry basket and a bottle of Tide with the caption "Mom's dead, gonna do some laundry."

I thought it was funny. I laughed, and that made me feel better. Do you know what the insane thing is, though? In response, people were actually telling me, "That's not funny." "You shouldn't joke about that, it's disrespectful."

The most absurd part, of course, is that these people actually saw themselves as the compassionate ones. They were the Good People, protecting standards of respect and decency by standing up to me, the Insensitive Jerk. Fancying yourself *compassionate* for criticizing someone's perfectly harmless coping method for their grief is twisted, to say the least, but it's also exactly what happens when we give decorum for decorum's sake a higher value than it deserves.

Of course, the fact that my mom was dead didn't matter to them. They were mad they had to think about dead moms for a second.

Whenever I think about our culture's relationship with death, dying, and grief (and why I think that mine is better), I think about season 8, episode 15 of *Seinfeld*. In case you, unlike me, were never broke and living in a run-down apartment alone in Long Beach, California, with no Internet, television, or friends and your only entertainment on the rare day when you weren't working was watching *Seinfeld* DVDs you had (to fight through drug addicts) to rent at the library on your barely working laptop that you could finally get to turn on again because the pilot of the plane you flew on for your traffic-reporting job took enough pity on how pathetic your life was to get it fixed for you for free—that would be the episode when George Costanza finds out that his girlfriend, Allison, is planning to break up with him, and he decides to (not) handle it by doing everything he can to avoid her, saying, "If she can't find me, she can't break up with me."

His thinking here is, of course, both absurd and hilarious. After all, the fact that George desperately hides to avoid hearing Allison actually say the words "It's over" doesn't make it any less over than it already is, right?

If you, like I did, laughed at that plotline—but can't laugh at things involving dying or death—you should ask yourself why.

Whenever I advocate for making jokes about death, a common response is "But death isn't funny!" And with all due respect? No shit.

Of course it's "not funny," but here's the thing: Are breakups "funny"? Have you just been consistently LOL-ing through all of yours? Happiest you've been in your life?

I mean, hey. Maybe you're one of those "lucky" (the use of "lucky" here, by the way, *does* mean "sociopathic") people who have never been hurt by a breakup. If so, allow me the honor of informing you that they can actually be quite painful. I have had some bad ones; I have had some brutal ones; I have even had some that have crossed the bridge from brutal into life-threatening.

See, going through a breakup kind of also qualifies as "not funny." People kill themselves over breakups; people kill *each other* over breakups. In fact, they do it so often that there are not just multiple episodes, but multiple whole-ass shows about lovers turning into murder-lovers when things go south. (If anyone who worked on Investigation Discovery's *Scorned: Love Kills* is reading this, quick question: Did the largest part of the show's budget go toward buying lingerie? That has to be the case, right? There is lots of sex on that show, and the people having it are always wearing intricate lingerie. Or do you only hire people who have their own massive lingerie collections already? People who just don't leave the house without their keys, phone, wallet, and garter belt? Do those people exist? Please help.)

Even if your breakups haven't been as bad as mine—and even mine were, thankfully, not as bad as the ones I've seen on Investigation Discovery—the reason we can laugh at plotlines and jokes about breakups isn't that they're inherently funny, but that there isn't anything that *can't* be funny in some way. Actually, I've found that the harder something is to talk about, the funnier the jokes about it can be.

Of course, some jokes or comments about death, dying, or grief miss the mark. It happens all the time, and it can hurt, but it's still been far from the greatest source of pain when it comes to our culture's communication on the subject. For me, one of the hardest things was the opposite: the expectation that I refer to my mom's death only in the most somber of ways, if I ever even dare to refer to it at all.

People like that exhaust me, and not just when it comes to something *serious* like my mom's untimely death. For example: I think about people being offended that I threw myself a funeral for my thirtieth birthday, entering the festivities by coming out of a casket in the back of Misfits bassist Jerry Only's sparkly purple hearse.

Like . . . okay? *You* didn't have to go. You weren't even invited, so you didn't even have to waste any finger power texting me to say that you couldn't make it. Why do you care so much? It was *my* birthday, and I wanted a birthday funeral. *I* wanted to arrive in a casket in the back of a hearse, and *I* wanted the hearse to have smoke machines in it (Jerry's idea, but I loved it), and *I* wanted to force my dad to give me a eulogy at a bar/sushi-and-steak restaurant while I was still very much alive.

A little weird to most, I guess, but so what? Why do you think that the virtuous thing to do was reach out to me, a complete stranger—tagged and all!—to cut me down for doing something that made me happy? Like, if you're "normal," that means that you probably spent thousands of dollars on bottle service at the TAO Downtown Nightclub for your birthday, but you don't see me claiming some kind of moral high ground over you because I made a different choice. Although honestly, I've got to say, I would much rather be trapped in a coffin (Jerry, by the way, made it for me, custom to my measurements; shout-out to his mom for sewing the pillow in so I'd have a comfortable place to rest my head on the ride there) than ever willingly subject myself to being trapped in a room surrounded by sweaty drunk horndog lunatics on blow who are all united in their elation that the DJ is playing whatever trash techno song sounds like all of the phones in a Verizon store started ringing at the same time.

To be fair, in some ways, I do kind of get it. It's not, after all, like I always talked about death the way I do now. Actually, part of the reason I'm able to do it is the example that my mom set with her own sense of humor about her dying.

There are too many examples of that to name them all, but to name a few: She was flirting with the male nurses up until the very end. When one was changing her bedpan, she assured him—in front of both me and my father—that, although her terminal illness had made her very thin, she used to have "a great ass." I heard some nurses in an elevator indicate her room as being the one with "that lady, the one with all of the pictures of the pope and the dick jokes." The night she died, a nurse in the room coughed, and my mom said, "Come on, man, you're going to get me sick!" The day before my mother died in a hospital in Boston, a woman who was always a pain in her ass at her job back in Michigan emailed her asking when she could come over to pick up some materials for an upcoming conference. My mom replied with the following:

> You won't have to bother, because nobody will be home.
> I will just be direct about this: I am preparing for my last days. I will not be returning home as a regular Delta passenger. I am ready to be received by the Lord. Love you.

She sat in her hospital bed—equipped with a million tubes *and* the full knowledge that she was about to die—and she was laughing her ass off. Rather than crumble under the weight of the realization of her impending doom, she decided she was not going to miss an opportunity to use it to troll *that* colleague. I respected it; I laughed, and I'll never forget it.

The biggest reason that I can speak frankly about (and even laugh at) death—bigger than my mom's example—is my having to endure the opposite: all of the people speaking in platitudes at me for the weeks afterward, because they wanted to be careful not to say anything accidentally untoward.

Dealing with someone you love dying is pretty hard, but a close second on earth has to be getting flooded with all of those bullshit "*She's in a better place!*" messages from everyone you have ever met right afterward. And the bereavement cards? My *God*. Who even *are* the people who write those, you know?

If you have had someone close to you die, let me ask you this question: Do you remember a single card that made you feel better after it happened? Do you remember what any of them said? Did you even bother to read any of them? I hope you didn't, because if you did, it probably just made you feel worse. When you are feeling devastated beyond hope, the last thing you want to look at is some kind of flowery bullshit, embossed in gold on a piece of parchment—as if whatever it's saying is supposed to be something that your barely there brain could even compute at that moment.

Why isn't there one that just says, "Fuck this, right?" and then there's just a cute lil' .5-milligram Xanax taped inside? Like, enough to take the edge off, but not so much that you'll forget who your great aunt Donna is in the receiving line? That would be actually, maybe, at least a little bit helpful.

Of course, the problem is that the people on Team In a Better Place think that they *are* helping, so it's hard to hate on it—at least out loud.

Hearing somebody put things all euphemistic and mild just won't fit how you feel. It will be so far from it, it will actually run the risk of pissing you off with its ignorance. It sucks, and you feel broken, and you just want someone to acknowledge that without trying to make it pretty.

No one does that. In fact, they do almost everything else *but* that. One of my favorites? "At least she didn't die in a car accident or something, and you got to say good-bye."

Like—okay, wow, great! You should be a grief counselor!

What I'd always say was, "I know," but what I always *wanted* to say was, "Do you go to people's funerals who *did* die from car accidents and tell their loved ones: 'At least she wasn't violently raped in front of you, and torn limb from limb, and set on fire while you watched'?"

Yeah, it could have been worse. Everything always could have been worse. Now shut up so I can focus on trying to remember who this lady in the receiving line is.

Here's the thing that all of those people freaking out about "what to say" need to know: There's nothing that you *can* say anyway, so please stop making it worse by thrusting *your* anxiety onto an interaction that I can promise is far more stressful for me than it is for you. I know you may feel weird about death or whatever, but I am getting hit in the face with it, so please don't add to that by making me also worry that you may be feeling uncomfortable.

Having to worry about making other people uncomfortable—ironically due to the same cultural expectations that were put in place to be *sensitive* and *protect* me—used to really, really freak me out. It made me second-guess everything I said and did.

Hell, it got so bad that it made me start second-guessing how I would handle things that hadn't even happened yet. I remember asking my sister how long after Mom died that people who asked us, "Hey, how are you?" still meant it in the somber, I-know-your-mom-just-died way. You know? If you are talking to someone in person, it's easier to tell. But over text? Are you checking in on how I am doing in terms of dealing with my traumatic life event, or do you just mean it as a greeting?

It's tough because if it's the former, you're expected to say something like, "Ugh . . . it's really hard sometimes, but it's great to have my family around," to which they will undoubtedly reply, "I am so sorry; I am here if you need anything." (Except without the semicolon; for some reason, most people never use those.) If it's the latter? Then, of course, you're supposed to say, "I'm good, hbu?!"

You would, after all, *not* want to misunderstand what they were going for. If they meant "How are you? Your mom just died, so I'm checking in," and you say, "Good!" then you're an asshole with no feelings, and they're going to start posting memes about sociopaths on their Instagram story that are no doubt directed at your callousness. On the other hand, if they meant "How are you!" in the nonchalant, "I'm-starting-a-convo-hiiiii" way, and you go on about how you have been crying all night, they're going to talk behind your back about how you're so unbelievably unstable that it's exhausting for them to continue to keep you in their life.

Then there was the issue of dealing with the people who did not yet know that I had a dead mom. The people who entered my life after that happened, the people who would assume that I *must* have a living mom, because I was barely twenty-six years old. Two years in a row, a *Fox & Friends* producer asked me if they could bring my mom to New York City for a Mother's Day segment. Each year, I called my friend Dave Navarro and laughed about what I could potentially say, because his mom had been murdered when he was a teenager, and him having the same sense of humor about tragedy was what had started our friendship in the first place.

I remember laughing as we threw around responses such as "Really? You could bring her? My prayers have been answered!" or "Are you sure? Because I think it might be harder than you think" or "I'm not sure hair and makeup could get her ready for air." I had the best time laughing with him about the possibilities—including his suggestion that I simply say, "I'm sorry, but I'm afraid my mother is unavailable . . . forever," without any other context. Then I'd eventually go back to reality and try to craft an email to producers that wouldn't result in an experience like the one in the makeup room after my multi-month My Dead Matriarchs' Funerals Tour.

By then I'd already done that laundry basket post, of course—but I'd also then subsequently *questioned* it. I judged myself for having impulsively hit "post," not least of all because of the emails I got telling me I was being offensive.

Then, one day, it dawned on me: I, too, was becoming obsessed with the flowery platitudes. Worse? I was doing it to make *other* people comfortable, not even to make *me* comfortable, even though *I was the one grieving.*

It is so unbelievably dumb that we do this to each other. How could we possibly be talking about the fact that someone *died*—which, by the way, also means *is dead forever*—and our minds shift to "Oh my God, there is *nothing* more crucial than making sure this is a comfortable conversation! I sure hope I do not use *this voice and breath that I still have the ability to use* to say something *offensive.* That would be the worst!"

The truth is, the only reason people treat this issue this way is that they feel like they have to in order to be compassionate and kind. Yet it just makes everyone in this situation feel a million times worse. It's uncomfortable for the people going through it; it's uncomfortable for the people trying to help the people who are going through it. It would help everyone if we could all just get real—so why don't we?

Now I do things differently, and I do so at every chance I get. For example, when I told my doorman last year that my dad was coming to visit and he asked me, "Is your mom coming?" I replied, simply, "She's dead," and then when he replied with that knee-jerk "I'm sorry" (which is, truly, nothing more than a reflex to try to get people to stop feeling ashamed or embarrassed of having made some kind of faux-pas than anything else), I replied the way I always do when that happens now:

"It's okay. You didn't kill her."

*And then we laughed.*

Now? Now I don't care if I make people uncomfortable. If I miss my mom? *I say so.* If I think of something from the past that bothered me about her? I say that, too. I don't bury it, deny it, or feel bad about it. I also love when other people tell me stories about her; I just wish they would stop awkwardly apologizing for having mentioned her when they do it. Memories and stories are all that I have left at this point, and it's not like bringing one up will suddenly make me realize that she's gone. Believe me, I have already noticed.

And most of all? If someone says to me, in an annoyed voice, "Ugh! My mom is calling me!" I love to say, "*Crazy! I can't get mine to call me back! Can we try your phone?*"

Then I laugh hysterically . . . and enjoy that flood of serotonin that my laughter brings me, with the added benefit of showing myself that even the scariest, saddest thing about life—the fact that it ends for all of us—isn't too scary for me to poke fun at it and laugh in its face.

CHAPTER 7

# IS THERE SEXISM IN COMEDY?

Is there sexism in comedy?

I don't know. I do remember this one night when I was terrified on a couch in a hotel room in Atlantic City, New Jersey, as a disgusting, self-tanner-abusing wretch of a man was hovering over me and trying to kiss me.

I'd met this waste of Jergens Natural Glow a few months before. He was the feature for the headliner I had sort of been working for sometimes and whom I was thrilled to be working for, even sort of and even sometimes. (Some of the gigs were paid in cash, others in The Exposure.)

So I'd taken the bus to Atlantic City out of the Port Authority terminal in Manhattan because Jergens had promised me an opportunity to be *his* feature. He told me that I could eventually make some dollars working at the club if the owner liked me. (I was pretty low on dollars in those days, if "bus out of Port Authority" didn't clue you in.) Even better, Jergens also said that it could help me with getting more work with that headliner. I was told that I could do one night, maybe two.

The bus experience was a nightmare, even for someone who had as much bus experience as I did. (Not to brag.) For example: While I was waiting at the stop, there was a guy with his pants down around his ankles, spinning in a circle and *pissing everywhere*, like one of those Home Depot revolving sprinklers. I managed to dodge it (seriously, sorry for all of the bragging). Then, on the bus, there were people smoking cigs (and not even sharing them with me!) and nothing to entertain me except for this phone call I overheard where a woman was arguing with her daughter about who was going to pay for her granddaughter's abortion. (The thesis of her argument? *She* had had to pay for the *last one*.)

Quick tip for any young women out there who might have to take a sketchy, late-night bus trip themselves: The best defense is a good offense. The crazy people won't bother you if they're too busy being concerned about how crazy *you* are. Look insane, and scratch as much and as vigorously as you can. I got so good at this, in fact, that I developed a motto: "If there is an empty seat on the bus, and it isn't next to you, then you didn't try hard enough."

I got to the hotel, and then—and *only* fucking then—did Jergens tell me that I would actually be staying with him in *his* room, because there were "no more rooms left." Yeah: the "condo" I was told we would be sharing was actually a room with just one bed.

I knew that the "no more rooms" thing was bullshit. There had to be other rooms there, and even if there weren't, there had to be some goddamn room nearby; this was not a Christmas Eve manger-birth situation. I knew that, but I also knew (and certainly, *he* also knew) that those facts didn't matter, because it's not like I (or anyone who would have ever even accepted an offer to take a bus to such a nightmare of a gig) had the resources to actually do anything about it.

I felt beyond uncomfortable and threatened and afraid, but also like my only choice was to just deal with it. It was really late, so I just sat on the couch and tried to relax—doing my best to convince myself that it wasn't a big deal and everything would be fine.

Still, my gut feeling about the whole thing really bothered me, and that lasted into the next day. I felt really creeped out, but did my best to just go downstairs and do my set anyway . . . which I obviously bombed. I was really preoccupied with worrying about what things might be like in the room later that night, especially since I kept feeling like he was looking at me weird. Sure, the night before had gone okay—but I'd also gotten in really late, and he'd just wanted to go back to bed. How might he behave when he wasn't half-asleep? I kept trying to convince myself that I was just overthinking things, but still couldn't shake the feeling that no one would deceive someone about sleeping arrangements if he didn't have some kind of nefarious motive. My head was spinning as it fought with itself, and I couldn't snap out of it. Plus, the audience was not really my vibe. To Jergens's credit, they absolutely loved the bit where he came out with a guitar and the speaker blasting "Save a Horse, Ride a Cowboy," putting the microphone in people's faces so they could sing along—only to have the music stop while the microphone was in someone's face, replaced suddenly by a loud sound clip that said, "I have a tiny wiener." Fucking *art*, man.

After the set, I decided to get hammered, because, well, I'm sure you have made a fuck-it-I'll-get-hammered decision before, right? When you know it's not the best idea, but you're so tormented by every idea in your head that you just want to make all of them go away for a little while? So I did, sitting as far away from him as the table would possibly allow, and listening to him telling me that he wasn't sure if I would get to do that next show or not. I remember feeling astounded at how obscenely orange his skin looked in the dark casino lighting, and wondering just how much time he must spend each week applying self-tanner—and, more importantly, *why*. I'm pale as hell, sure, but I've done fine just rolling with that. I don't have the kind of patience to paint myself orange, and the sun? Forget it. I mean, I'm Polish! We've evolved to survive long periods in dark basements evading capture, *not* to survive in the sun.

So that's how I wound up on that couch, having to block his disgusting, aggressive attempt at hooking up with me—as if purposely placing myself on the couch hadn't been a clear enough sign that I'd rather fling myself out the window than fuck him. Like, *Jesus, Jergens, I am on this couch for a reason, please go away, you resting-duck-face-ass creep.* I remember waking up in the morning and being grateful that my clothes were still on when he came in and told me that I would, in fact, *not* be doing the second show. He, of course, made it sound like this was because I didn't have a good set. I totally would have believed that if he had told me that after I didn't do well, instead of waiting to tell me twelve hours later after I didn't do . . . him.

In any case, I felt relieved. No, I didn't want to miss out on an opportunity, especially since my life at that time was extremely devoid of those, but I was still so, so, so relieved. I rushed out and spent most of my time waiting for the bus, crying next to a slot machine. It was the best I'd felt in at least thirty-six hours.

Note: That next morning, Jergens had also told me that it was really great that we didn't hook up, because that would have been "unprofessional." Yes, seriously. All these years later, and the thought of him still makes me want to throw up.

Later, I would find out that Jergens had talked a bunch of shit about me to that headliner at their next gig, and that his doing so was the reason that I never got to work for that headliner again. (If you've been trying to figure out who "Jergens" is, just stop. You will never guess, because you have never heard of him.)

But other than that? No, there is no sexism in comedy. Everyone is looking for a funny chick! If anything, women have it easier because so many people *want* a funny chick, and it's just so unfortunate that so many of them aren't.

Look, I know that it may feel a little crazy for me to be talking about sexism in the first place. Sexism seems so 2015; it's all about transphobia now. But topic trends, unfortunately, don't mean that sexism stopped existing. Experiences like the one I just described are, after all, extremely common, and the one I just described is far from the worst one out there. I know that because, well, it's far from the worst one that has happened to *me*.

It's also important to recognize that, whenever we do talk about sexual harassment and assault, we're usually having a conversation that's missing a lot of information. For every story that we do hear, there are many more that we never will.

There are a few reasons for this, but an obvious one would be nondisclosure agreements, particularly when it comes to the victims of powerful men. Those men *are* called predators, after all, because they seek prey—and, just like all predators, they purposely choose victims whom they see as too weak to overpower them. They do the worst things to the women who have the fewest resources, specifically because of their lack of resources. If one of those women actually does report the guy, getting a settlement and signing an NDA, then she likely still won't have the means to break her silence about what happened even if she wants to in the future. Yes, we've had the chance to hear stories from women who have become rich and powerful enough to share them without fear of legal repercussions, but there are countless more women who are silenced by settlements. Although NDAs can have their benefits—even for some victims, who may see the prohibition of talking about the abuse as a positive of the agreement, which I will get into later—they also inherently prevent the people who (unfortunately) know the most about sexual harassment and assault from being able to fully and honestly contribute to the conversations about it, even if they do decide that they want to at some point down the line. Sometimes, when you ask a woman to talk about this issue, you're asking her to do something that she legally can't do. Sometimes, when you shame a woman for not speaking out about a certain situation, you're shaming her for not doing something that she legally can't do.

(OK, so, quick break: If this were Twitter, I wouldn't have made it all the way to this paragraph without a Reply Guy piping in to tell me, "It happens to men, too." So, just want to say: Yeah. I know that. It does. I'm not discounting those experiences, or saying that they're not awful, because, of course, they are. It's awful anytime it happens, but it's also true that it happens to women more often, *and* that women are the subject of this chapter, and that I'm allowed to pick what I talk about in my chapters, because this is my book. You can write one, too.)

Plus, even aside from women who don't talk because of NDAs, there are all of the other women who don't have NDAs because they didn't want to talk. Actually, this describes most victims. As a 2017 article in the *New York Times* states:

> *[O]nly a quarter to a third of people who have been harassed at work report it to a supervisor or union representative, and 2 percent to 13 percent file a formal complaint, according to a meta-analysis of studies by Lilia Cortina of the University of Michigan and Jennifer Berdahl of the University of British Columbia Sauder School of Business.*

As the *Times*—and/or, even a moderate amount of listening to people's experiences and thinking critically about the issue—explains, there are several reasons for this. For one thing, as I've already alluded to, victims of the more powerful and better connected often think that there's no way that their story could win when matched up against the power and connections of those perpetrators. But there's another reason, too—and it's actually the same reason why the NDA arrangement may be attractive to those who do report: Once you go public about being a victim of sexual harassment or assault, especially if it involves a high-profile perpetrator, then that tends to become the main thing that you're known for. This is especially true if you're at the beginning of your career, which, again, is often the exact profile of a woman that a predator will strike.

Jane Park, who worked in "business consulting and strategy" before starting her own beauty company, Julep, told the *Times* that she never reported any of the misconduct she experienced during those years for that very reason: "It's made into such a big deal that you have to make a decision: Do you want to ruin your career? Do you want this to be everything that you end up being about?"

Now, you may notice that the issues I am discussing here aren't really ones that are unique to comedy, right? It's less a matter of Women in Comedy, and more a matter of, like, "Woman in Job," or "Woman Being a Person."

Or, as Amy Schumer put it in her 2012 special, *Mostly Sex Stuff*: "What's the hardest part about being a female comedian? The rape."

*Of course* sexual harassment, abuse, and assault are not unique to Women in Comedy; they're universal among women on all life paths. Although some stories are certainly more severe than others, I'm willing to bet that every woman could offer some example of facing harassment or sexism or mistreatment at some point during their lives—so why, then, are Women in Comedy so often asked about what it's like to Be a Woman in Comedy specifically?

Jen Kirkman, one of my favorite comedians, whom I've been following for years but who does not know that I exist, once said:

> *This question is the hardest part. It's yet again another opportunity for guys to say that I'm complaining or to retread the same old*

*stories. There is sexism in the world, so of course it bleeds into every single area of life. I don't answer this particular question anymore.*

Kirkman, of course, is far from the only Female Human Comedian to have this point of view. So many others have expressed similar sentiments:

> *What is it like to be a woman in comedy? I would say it's 1% jokes & 99% answering this question.*
>
> —Aparna Nancherla

> *I find it annoying that funny women always have to talk about being a funny woman. I'm a funny person. We're not charity cases. We're talented. It's done.*
>
> —Michelle Collins

> *I don't want to validate that stupid-ass question. People get so hung on gender, sexuality, and race, and they don't see you as a creative as they might, say, Jerry Seinfeld.*
>
> —Phoebe Robinson

I totally agree with all of this. Because, after all, like Kirkman said, female comics are obviously going to have stories about sexism—but that's because they are women, and sexism is a thing. We do live in a patriarchy, which is less of a political statement and more of a fact when you accept the fact that one definition of the word "patriarchy" is "control by men of a disproportionately large share of power," which is just objectively true. There are, statistically, more men than women in positions of power. (I honestly always love when some dude will get all weird and combative when I mention that we live in a patriarchy while he also considers himself to be a Numbers Guy.)

Of course, if you *are* someone who considers numbers, you'll also notice that there are currently more (at least well-known) male comedians than female ones. Just a matter of decades ago, actually, being a person who did stand-up comedy who was also a woman was somewhere between rare and nonexistent. Joan Rivers gets credited as the first woman to perform stand-up comedy in the current "confessional" iteration of the art form. As a piece in *Time* explains:

> *It was a new idea about how comedy should be presented: that it should be real and personal. The standups of this era would forever change how Americans viewed comedy. Woody Allen was emblematic of this movement. But so was Joan Rivers, the lone female standup of this new guard.*
>
> *At the time that Joan Rivers started, female standups were rare. The only other major comic working at that time was Phyllis Diller, famous for her rapid fire jokes about a fictional husband named "Fang." It wasn't easy for a woman to get up on stage and just talk. And were it not for Joan Rivers' fierce determination—there were many naysayers—she might not have made it.*

Not only were there naysayers, but a lot of the naysaying was explicitly focused on her gender. A *New York Times* review of Rivers published in 1965 stated:

> *Joan Rivers, a new comedienne of ripening promise, who opened at midweek for a two-week run at the Bitter End, is an unusually bright girl who is overcoming the handicap of a woman comic, looks pretty and blonde and bright and yet manages to make people laugh.*

Got that? "*The handicap of a woman comic!*" Like, holy shit, right? It is perhaps because of tidbits exactly like this that people are so interested in the whole Women in Comedy conversation, and, to an extent, I can sort of get that. But really only sort of, and actually more like "hardly." When you think about it, Rivers wasn't necessarily facing sexism because she was a Woman in Comedy, so much as because she was a Woman in 1965. Like, what was life like back then for all of the women who were *not* doing comedy, you know? Back then, banks could (and did) refuse to give credit cards to unmarried women, because they could require a husband's signature to open one. (This requirement didn't become illegal until nearly a decade later, with the passage of the Equal Credit Opportunity Act of 1974.) In several states, women couldn't serve on juries. Yale and Princeton did not accept female students. It wasn't until June 1965 that the Supreme Court ruled it illegal for the government to deny birth control to unmarried women. Just three years earlier, President John F. Kennedy had said in a broadcast conversation with Eleanor Roosevelt, "We want to be sure that women are used as effectively as they can to provide a better life for our people, in addition to meeting their primary responsibility, which is in the home."

(I've said this before, but I'll say it again: As a woman who still routinely stores her clean laundry in the dryer to avoid having to actually put it away, I would have been totally screwed living in a time before it was cool for women to have careers.)

To me, the more interesting and impactful conversation is less about Women in Comedy specifically, and more about women in general—and how whatever Women in Comedy may be going through reflects what women are going through everywhere.

Maybe it's just because I'm a woman and have gotten my period before, and that makes me more dumb, but I've always heard a pretty clear implicit sexism in the question: "What's it like to be a Woman in Comedy?"

In asking that question, after all, you're either suggesting that being a Woman in Comedy is unthinkable, or, at the very least—whatever it is you may be doing in comedy—the fact that you are doing it *while also being a woman* is the most noteworthy aspect to explore, rather than anything about your unique career as an individual artist.

Every single one of the Women in Comedy whom I have referred to in this chapter has, throughout her career, been lumped in countless times with the group of "female comedians" just because they are women, as have all female comedians—even though, as entertainers or as people, they may not really have all that much in common.

For example, Amy Schumer is a white person who was born into a "really wealthy" family on the Upper East Side of Manhattan that went bankrupt by the time she was nine, the same age she was when her father was diagnosed with multiple sclerosis. Schumer's parents got divorced when she was a kid, and her first breakthrough came when she won *Last Comic Standing* in 2007. She's now married with a child.

Phoebe Robinson is a black person from the suburbs of Cleveland who got her start as a writer on MTV's *Girl Code*, and who has said that she initially "really could not have cared less about stand-up." Her parents are still together (at least according to the most recent information I could find), and her big break came from starting a podcast with her friend called *2 Dope Queens*. As of this writing, she is not married and has no children.

If it's still not clear how different they are, just try to imagine Schumer having named her first solo stand-up HBO special what Robinson named hers: *Sorry, Harriet Tubman*. They are different as people, they're different in terms of their material, and I think it's annoying and stupid to group them together just because of their chromosomes.

Honestly, I also hate when people think they are championing

Women in Comedy by having what, to me, is quite clearly a patronizing attitude toward them. For example: No, I don't think that you're a sexist if you don't find a particular woman funny, but I actually *do* think it's sexist to insist that the only reason someone might not like a particular comedian would be because that comedian happens to be a woman. Sexism is an issue, yes, but that doesn't automatically mean that every issue a woman faces is a matter of sexism. This, of course, hurts a lot more than it helps. If everything is sexist, then nothing is. If you use it to describe everything, people will be less inclined to hear you out even when it's real. Plus, if you are putting a female comedian on a show simply because you feel like you need to slap a chick (any chick!) on the bill—and not because you think she's funny enough to be there—then you're actually helping to spread the absurd, tired Women Aren't Funny trope, rather than the opposite. To me the dream is to be evaluated only as a person—based on my talents, abilities, and other attributes—and treated the same way as people would treat anyone else.

Actually, this is the exact reason why I think it's so great that Gutfeld roasts me on the show so much. In case you haven't noticed, Gutfeld makes fun of people. If he held back on making fun of me because of my gender, then I'd have to interpret that as him believing I was less able to handle what Brian Kilmeade can handle just because I'm a chick. And what could be more insulting than that? Because the truth is, not only can I "handle" it, but even the *Atlantic* admitted that I can also "gamely field" it. Actually, I can do that so well that I've spent years getting paid for it. Speaking of money, I never would have gotten this job if Greg didn't feel like he could make fun of me (seriously, he needs to tease people in order to function) and so if you're going to reach out to me or him and say that he should be "nicer," please understand that what you're really saying is that I should be fired . . . which isn't very nice at all.

I do, by the way, get that sort of feedback from people. Both of us do. Sometimes the response to one of his jokes will be, "Be nice to Kat!" or "You're too mean to Kat!" I hate it for all of the reasons that I just explained, but also for one more: It assumes that I myself did not write whatever joke about me that you're getting offended by on my behalf, which isn't always the case. Sometimes I *do* write them.

A few examples, just for fun:

*"She's like a syringe . . . sharp, skinny, and full of medication."*

*"She's like a toothpick . . . skinny, sharp, and easily fits into dark, disgusting places."*

*"She's like an alarm clock . . . will not stop making noise until you pay attention to her."*

*"She's like a praying mantis . . . skinny, bright, and may kill and eat her husband."*

*"Her marriage is like the milk in my fridge . . . only a week old and already going bad."*

*"Every day her husband tells her three little words: 'What've I done?!'"*

*"She's like glitter: bright, fun, and impossible to get rid of."*

*"She's like a rubber band . . . thin, useful, but bound to snap sooner or later."*

There are countless more, but you get the idea. I wrote these jokes, Greg used them to introduce me on the show, and then, for at least some of them, he got crap for being too mean to me. Like . . . uh, I guess that means you think that *I* was too mean to me? Which is sometimes true, but if you think that these things are too mean,

I'd really hate for you to have to hear my inner monologue. Bitch, you wouldn't last a day in my head. But that's more a conversation for me to (continue to) have with my therapist.

The truth is, Joan Rivers wasn't a funny woman so much as she was a funny person—and actually not just funny, but uniquely so. When I say "unique," by the way, I mean "irreplaceable." I mean it so much that I still remember tweeting in October 2016: I would kill (personally murder) any one of you to have Joan Rivers around for this election. It wasn't because she was a "pioneer" for women that I hated losing her, or because her cultural significance had made her some kind of relic, but because she was *her.* Some people were certainly pissed off by that tweet, but I'm equally certain that I'm not sorry for it. Not only because it was a joke and people should calm down, but also because what I expressed through it has been an enduring feeling. I have had that exact same thought at countless moments as the years and absurdity have continued. It's a shame we can't hear what she would have to say about so many things, not because of who she was in terms of The Culture, but because what she would have said would have been awesome.

It should be no surprise to anyone that Rivers herself actually hated discussion about her being a "pioneer" while she was still alive. In her interview on the PBS series *Pioneers of Television* that aired in 2013, Rivers actually pushed back on being described by the show's title, saying:

> *It upsets me to say "I'm a pioneer," because I am so current now, do you know? I get very, I don't, when the ladies come up and say, "Oh, you broke barriers for women," and I go, "I'm still breaking barriers," that's starting with it, and I can still take you, sweetheart, with both hands tied behind my back. You asked me . . . "Am I proud to be a pioneer?" I'm not a pioneer. I'm still in the trenches, I'm still breaking ground, I have never spent two minutes saying,*

> *"Well, I just did that," I'm still looking for the new frontier. I'm still in my astronaut suit.*

When Rivers died the year after this interview aired, I wrote a column for *National Review* titled "Doing Feminism," discussing the resulting debate about whether Rivers had counted as a feminist or not:

> *Some said no—she said offensive things about other women's bodies, and that's not feminist. Others said yes—she may have said offensive things, but she did so without caring what anyone thought, and that is feminist. And all of them had the wrong idea.*
>
> *In terms of the first group: Sure, Rivers was ruthless. She called a different woman fat almost every day and even threatened to charge HBO with crimes against humanity for showing Lena Dunham's fat naked body on television too often.*
>
> *Demanding that another woman limit her artistic potential isn't exactly feminist. Neither is suggesting that Dunham or any of the other female targets of her jokes would be too fragile to handle this kind of criticism just because they're women.*
>
> *But here's the kicker: The pieces on the opposite side aren't any better. The authors who praise Rivers because she was a "feminist hero" who "paved the way for women in comedy" are well-intentioned, but they miss the point.*
>
> *Rivers wasn't a woman in comedy. She was a comedian in comedy.*
>
> *"I don't help women because they're women; I help whoever's good. If you're good, you're good, and deserve to get ahead," Rivers declared in 1983.*
>
> *The fact that so many people focused on whether she was or was not a "feminist" misses the entire point of what an actual feminist society would look like.*

*It's like when comedy clubs host "all-female showcases" with names like "She-larious!" or "Her-larious!" or "She's Her-larious!" as though that's somehow empowering women—when empowering would be an all-female show without the need to call it that.*

*Yes, being a female comic comes with its own particular set of challenges. All too often, men in the crowd are more interested in sexually harassing you than listening to your jokes. I've purposely messed up my own hair or taken off my makeup before going on stage to try to avoid it. I'm worried people will think I'm "too pretty to be funny."*

*And Joan Rivers would have shamed me for that. Not because my behavior "fueled the fire of the patriarchy" or some other weird "feminist" trope, but because it simply wasn't worth my while.*

*"I didn't have time to go up to anyone and say, 'Go out and make it in a man's world,'" she said in an interview with* Playboy *in the '80s. "I just said, 'Look at me and you can see what I'm doing.'"*

*"Doing." That word is what's missing when I think about what "feminism" has become today. It's a lot less about what women are doing, and a lot more about what society is doing to women.*

*The Internet is filled with blog posts from so-called feminists seeking to expose new ways that our culture has been secretly oppressing women. Disney films hurt little girls by chaining them to traditional gender roles instead of telling them they can grow up and have careers. Common phrases such as "Oh, man!" and "Hey, guys!" hurt women because they're constant reminders that we live in a male-focused society, and that's discouraging.*

*Since when is looking for ways to be a victim empowering? It's not. It's the opposite.*

*So, was Joan Rivers a feminist? Whether you say yes or no, she wouldn't care. She was too focused on doing what she wanted to do as an individual. And that's the most empowering stance of all.*

I may have been twenty-five years old, but damn, I was right. Yeah, there's stuff I deal with as part of my job that men with the same job don't have to deal with (like the Facebook comments telling me I shouldn't speak until Greg addresses me and asks me to speak first; LMAO, who hurt you, Bernice?!) and it's fine to talk about those things . . . but only if you realize that it's not a Woman in Comedy Thing so much as it's a Woman Thing. I might have to hear more opinions from strangers than most people, but really, the conversation to have about the sexist ones should be about how that sexism exists and not about What It Is Like for Me as a Woman Who Is in Comedy. The bottom line is, even though I may be a woman, I am also a person—and a person who would never, ever, ever fuck a Jergens for a job.

CHAPTER 8

# WORDS ARE NOT VIOLENCE

A lot of people complain that awards shows are too boring these days, and they usually are. The 2022 Oscars had just 15.36 million Americans tuning in and was *almost* as boring as the previous year's, which had the lowest-record rating viewership ever.

Until, of course, The Slap.

The Academy Awards are supposed to be a fun celebration of great films. Now it's just a strained demonstration of wokeness, a time for celebrities to show everyone watching at home that not only are they richer and more successful than you, but they're also more enlightened. It's become the entire focus, and it's only going to get worse. The Academy of Motion Picture Arts and Sciences has already announced that beginning in 2024, directors and producers will have to submit details about the gender, race, disability status, and sexual orientation of their films' casts and crew and will be disqualified if they don't meet certain quotas. To me, that's pretty creepy. Say you're just some guy looking to build a set or do the sound—and then you find out that you have to fill out a form about who you like to bang first. I mean, the intake forms for a gastroenterologist are less invasive!

Obviously, it is unconscionably cruel to reject people because of their sexual orientation. But it is also cruel to pressure, let alone to *expect*, people to reveal details about it on anything but their own terms. In the fall of 2022, actor Kit Connor was accused of "queerbaiting" for playing a gay character on Netflix but holding hands with a female costar in real life. The bullying got so bad that he ultimately ended up tweeting: "Back for a minute. I'm bi. Congrats for forcing an 18-year-old to out himself. I think some of you missed the point of the show. Bye."

Sometimes people might not feel safe coming out. Sometimes people might not feel comfortable, because it just seems too personal. Sometimes people might be struggling to define their sexuality to themselves, and therefore feel they'd have no idea how to begin to explain it to others. Sometimes people might feel totally cool telling some people, or even most people, but would rather not make any sort of public or otherwise official announcement. Sometimes it's something else. But *all* of the times? It's no one's fucking business unless the person in question wants it to be.

When did things change from "What you do in your bedroom should be no one's business but your own" to "Our business requires that you tell me what you like to do in your bedroom. Write it on this document, which I will then submit to a third party!"? As yucky as that announcement felt, it also wasn't all that surprising. This was, after all, the same industry where directors had already been getting shamed for, say, casting straight people to play gay characters—which, when you really break it down, is telling those directors that what they *should* have done is said, "We're sorry, we would love to give you this role, but we're afraid that you're just not fucking the right people." Like, ooooookay, Harvey Weinstein.

Anyway, it's not that diversity itself is boring—it's so not boring, actually, that it makes Hollywood's ability to make it boring almost kind of impressive. It *shouldn't* be boring to learn about people and cultures and experiences that are different from yours. It's just that, when it's so contrived and performative rather than honest, curious, and open, people know it.

I actually didn't mind the Oscars in 2022 being a snoozefest, because that's just what I was about to do. I'd snuggled into bed by 10:30 p.m., excited about how much rest I was going to get, planning on working out before work, so proud of myself for being such a responsible adult, so far removed from the clamorous, Prosecco-guzzling Mimosa Monster that I'd been on Sundays in my twenties. Even if I didn't drift off until 11, I could still wake up at 7 a.m. and get eight hours of sleep! (Yes, the only thing more boring than the Oscars was me.)

But then Will Smith slapped Chris Rock. He ran up and hit him right in the face, using physical violence to punish Rock for making a joke that referenced Smith's bald wife, Jada Pinkett Smith—or, put another way, for daring to do the job he was literally hired to do: making fun of the celebrities in the crowd. It got even messier when Smith, who was able to just stay there and chill after assaulting someone, won Best Actor for *King Richard* (which, until then, I had assumed was a film adaptation of some sixteenth-century play) and got a standing ovation.

Bedtime was a bust.

It was shocking, no doubt. But nowhere near as shocking as the response in the wake of it: Smith had assaulted a comedian *over a joke*, and people were *defending* him for doing so.

A YouGov survey of 1,319 Americans found that 61 percent of respondents thought Smith was "wrong to hit Chris Rock after his joke," with 21 percent saying that he wasn't, 19 percent saying that they weren't sure, and only 59 percent saying it is "not ever OK to hit someone for something they said."

Can you believe that it wasn't even close to 100 percent of people categorically condemning a physical assault over a joke? It gets worse. A poll conducted by Blue Rose, with 2,162 online responses, asked the question in a different way: "Which side was *more* wrong?" (emphasis mine), and 52.3 percent said that *Chris Rock* was.

Sorry, but that is unhinged. Completely, indefensibly lunatic-level crazy.

I understand, of course, why people might think that Rock's joke sucked. Not only was it hack, but alopecia is also devastating. As you may recall, I myself was diagnosed with it in my twenties after experiencing what my doctors called "severe" hair loss. I remember noticing that it wasn't growing how it used to, and the exact day that I knew I might need to get help. Although it was always thin, I used to be able to grow my hair down past my shoulders. It always looked like shit, sure, but at least I felt like I looked noticeably female, and I liked that. Especially since my mother had all but forced me to spend most of my childhood with a bob thanks to the aforementioned "looked like shit"—apparently preferring that I looked androgynous but kempt.

At some point it got really damaged, so a hairdresser cut a bunch of it off, which was fine . . . until it didn't grow back. It was so thin and short that, unless I was at work or coming from work wearing pounds of extensions, I would never, ever be seen with it down—choosing to keep it only in an ever-smaller messy bun on the top of my head, fluffing it out as much as possible, because I felt like doing so at least made the state of it look ambiguous. One day, I remember looking in the mirror after another hairdresser visit and realized that, even *with* professional styling, it was so pathetic-looking that I could still not wear it down, not at all, not anywhere, not even in front of my own mirror. I remember the sight; I remember the feeling; I even remember the shirt I had on when I looked at my reflection.

So, the next time I was at the dermatologist getting a cyst injected on my face (I swear all I do is hot girl shit), I decided to bring it up. I guess I expected they'd probably tell me it was going to be fine, that it would get better on its own. I mean, it would hardly be the first time I'd worried needlessly about something, and it's not like I could be going bald at such a young age. Especially because I was, you know, a *girl*. It was going to grow back. This was a weird phase. I would have the hair again, and certainly not lose any more. I was, I guess, unable to comprehend the opposite.

But, of course, it *was* the opposite. The doctor confirmed that I had, in fact, lost a "severe" amount of hair, and they couldn't assure me that it would grow back, or that it wouldn't get even worse. They suggested I get my thyroid checked, and I prayed to a God I don't believe in that it was "just" my thyroid—"just" the organ involved in regulating my metabolic rate, muscle and digestive function, brain development, and bones—and not my hair, which is involved in nothing, other than my self-worth. (And, apparently, my worth in the eyes of this one guy I went on a date with, who told me—upon hearing that I was wearing hair extensions—that he was disappointed because long, healthy hair was a sign of fertility.)

Honestly, that visit to the dermatologist was even worse than the time they suggested that the bites all over my body, which I believed to be a recurrence of the shingles I'd just had (seriously, guys, *so* much hot girl shit), looked more like bedbug bites. Spoiler alert: They were bedbug bites. (Word to the wise: Never let a loser guy whom you don't even like live with you for months on end and cry at your apartment all the time and judge you for your workplace even though you're not together and he sucks and you have zero interest in anything he has to say because you don't know how to tell people no. He will insist on calling himself an actor, even though he has never made a dollar doing acting. Then, when he gets fired from the waiter job where he *had* been making dollars, you will end up paying for his acting classes because you made the wrong face after he said he would never take a gig on *Investigation Discovery* because it would ruin his reputation as a Serious Actor—despite, again, never having made an acting *dollar*, let alone an acting *reputation*. He sucks. I know it may be really hard to get rid of him, and that all of the normal tricks won't work, and that even your clearest demands will be contested and sometimes even met with a "hurt foot" that allegedly prevents him from going anywhere. I know that he will go to absurd lengths to avoid giving up his access to in-unit laundry, but do not let him get away with any of it. He will steal your time and spirit and give you nothing but bedbugs in return. I am sure that this is useful advice, as I'm confident that this is a common, normal, relatable problem.)

Anyway . . . unfortunately, at least in my view, subsequent blood tests revealed that it *wasn't* my thyroid causing my hair loss. It *was* alopecia. I remember having some friends over one day and opening up to them about it, taking my hair out of the messy bun. When I did, they all said the same thing: "We didn't realize it was this bad." (Aha! So the Messy Bun Trick *had* been working.)

I tried to cheer myself up the best I could. I tried to tune it out when the abusive Nightmare of a Boyfriend I had been "seeing" at the time—you know, that on-and-off, roller-coaster-to-hell kind of "seeing," the only kind of "seeing" that's possible when it comes to someone so volatile and controlling—would use it against me. ("*Fuck you, you rapidly aging idiot!*") I tried to tell myself that if it got worse, I could always just be, like, a fun wig girl. Hell, I could even get purple ones, and Fox couldn't say anything to me about having purple hair because I had a disorder!

But honestly? I felt awful.

I took all of the action I could. After my diagnosis, I started that regimen I mentioned earlier, which I still continue to this day: Rogaine every single night, supplements twice per day, a laser helmet every other day, and PRP, which stands for platelet-rich plasma. By the way, if you don't know what PRP is, let me explain it to you: First they take your blood. Then they spin it into a centrifuge to separate the platelets. Then they take that platelet concentration and *inject it all over your skull.* It's painful, it's expensive as hell, and they tell you straight up that they don't know if it will work for you. You may not know for up to six months of treatment.

Thankfully, from a combination of these things—plus, I assume, finally being rid of the Nightmare of a Boyfriend and all the stress that came with him—it is so, so much better now. I can finally grow it past my shoulders. Thin as hell, yes. Still wearing pounds of extensions at work, yes. Still terrified about losing it in the future, yes. But it is, for now, back to how it was before my diagnosis.

All of this to say, *I feel you, Jada.* No, I didn't go completely bald, so I can't say I know exactly how she feels, but I can pretty much guarantee you that if I did go bald, my skull shape would not allow me to look anywhere near as good as she does. Alopecia is upsetting. It sucks. It's heartbreaking. In a way, it's a loss of some form of self-identity. It makes you vulnerable in a weird way. So I can acknowledge how Jada felt when Rock made the piss-poor *G.I. Jane* joke. I have the fear that I, too, will one day have to shave my head.

The thing is, though, I'm less afraid of losing all of my hair than I am of living in a world where a comedian has to worry not only about getting hit for making a joke, but also about people believing *he* was the one who'd behaved *worse.* I would call it "unbelievable" if progressives hadn't already coined a phrase to justify this exact sort of twisted mentality: "Words are violence."

In 2017, as a columnist for *National Review,* where I wrote mostly about speech issues and our politically correct culture, I noticed the phrase popping up everywhere. At the time, a survey conducted by McLaughlin & Associates for Yale's William F. Buckley Jr. Program found that 81 percent of respondents believed that "words can be a form of violence," and 30 percent said "physical violence can be justified to prevent someone from using hate speech or making racially charged comments."

The same year, an esteemed emotion researcher at Northeastern University, Lisa Feldman Barrett, wrote a column in the *New York Times* called "When Is Speech Violence?" In the piece, Barrett claims that, since prolonged stress can damage us physically, it is fair to say that hurtful words can count as violence.

Jonathan Haidt, a social psychologist and professor of ethical leadership at New York University Stern School of Business, and Greg Lukianoff, the president and chief executive officer of the pro–free speech group Foundation for Individual Rights and Expression (FIRE), cowrote a piece in the *Atlantic* refuting some of Barrett's arguments. For example, they explained:

> *Feldman Barrett used these empirical findings to advance a syllogism: "If words can cause stress, and if prolonged stress can cause physical harm, then it seems that speech—at least certain types of speech—can be a form of violence." It is logically true that if A can cause B and B can cause C, then A can cause C. But following this logic, the resulting inference should be merely that words can cause physical harm, not that words are violence. If you're not convinced, just re-run the syllogism starting with "gossiping about a rival," for example, or "giving one's students a lot of homework." Both practices can cause prolonged stress to others, but that doesn't turn them into forms of violence.*

It's a great point, really. Using Barrett's logic, routinely eating fast food would be violence, as would routinely sitting on the couch instead of going to the gym. My behavior in the months following the end of my relationship with Nightmare—and particularly on a trip to Los Angeles that I'd taken to get over him—was *definitely* violence, and probably even a felony.

To Barrett's credit, her piece actually did make the distinction between long-term and short-term stressors. She wrote:

> *Offensiveness is not bad for your body and brain. Your nervous system evolved to withstand periodic bouts of stress, such as fleeing from a tiger, taking a punch or encountering an odious idea in a university lecture. Entertaining someone else's distasteful perspective can be educational. . . . When you're forced to engage a position you strongly disagree with, you learn something about the other perspective as well as your own. The process feels unpleasant, but it's a good kind of stress—temporary and not harmful to your body—and you reap the longer-term benefits of learning.*

The *Atlantic* piece gave her at least partial credit for this—agreeing that she was correct, but also adding that she "could have gone a step further": The sort of experiences she was referring to were not only "not harmful," but also potentially able to make a person "stronger," possibly prompting them to have "a milder stress response in the future . . . because her coping repertoire has grown."

I agree with Haidt and Lukianoff there, as well as with their added pushback on Barrett's logic in concluding that professional provocateur (and all-around asshole) Milo Yiannopoulos should, in fact, not have been allowed to speak at the University of California, Berkeley, because of his role in leading what Barrett called a "campaign of abuse." Although Yiannopoulos more than qualifies as a Total Dick, if her whole point was to distinguish between short-term and long-term stressors, it seems ridiculous to conclude that a speech should be forbidden when that speech itself would have lasted only a few hours maximum. Plus, of course, it's not like anyone was forcing people to attend. Apparently you can now be a victim of a Violent Words Attack without ever even hearing those words yourself.

Since 2017, things have only gotten worse. Then, people like Lisa Feldman Barrett could at least acknowledge the *potential* usefulness of exposure to offensive speech. But fast-forward to 2022 and you get a piece chock-full of quotations from "experts" in the fucking Health and Wellness section of *USA Today* about The Slap (titled "Will Smith, Chris Rock, and When Words Are Violent, Too") that was far less measured in its criticism of free speech—essentially arguing that, even after all of these years and cancellations, we are still not going far enough in terms of policing speech.

For example:

> *"Kids used to say, 'sticks and stones may break my bones, but words will never hurt me,' but we know that words do hurt, which is why it becomes incredibly important even in comedy to be thoughtful." Dr. Alisha Moreland-Capuia, director of McLean Hospital's Institute for Trauma-Informed Systems Change in Massachusetts.*

> *"Chris Rock got slapped pretty hard, but he probably can't feel that slap anymore. The sting was probably gone a few hours later. To the extent that there's lasting harm on his end, it's probably because of what's being said about him. It's not the physical act itself and that is going to be true for what he did as well—for Jada Pinkett Smith, the pain of that emotional violence is probably going to last for years." Sherry Hamby, founding editor of the American Psychological Association journal* Psychology of Violence.

> *"I would put that joke on the continuum of linguistic violence. We make too light of words. . . . A lesson that we can take from what*

*happened is the need to be more reflective about what we're saying and the harms that it can cause." William Gay, a UNC Charlotte professor who studies the philosophy of language.*

We make "too light of words"? Are you fucking kidding me?

In another quotation included in the piece, Hamby acknowledged that Smith would have been better off to use his own words than to hit Rock "because that probably would have taken Chris Rock to task and in a way that would've garnered a lot more sympathy and made the conversation much more about his bullying than about Will Smith's violence."

Although I agree that words would have been more effective, to call what Rock did "bullying" is absolutely nuts. He wasn't *bullying*, he was *joking*, and the fact that he *was* joking could not have been more clear. He was a comedian holding a microphone, literally and explicitly there to make jokes at the expense of the celebrities in attendance. What's more, those celebrities were all well aware that this was the arrangement, especially veteran A-listers like Smith. Comedians are supposed to push boundaries; it is part of their jobs, and it allows for comedy to be a potential vehicle for connection and healing, which is something I discuss at length elsewhere in this book. You can't be afraid to take risks, let alone worry about every possible situation that every person in the audience might be facing. When one Twitter user asked why Rock had referred to his joke "as a 'GI Jane joke' instead of 'a joke about a woman with alopecia'" right after Smith slapped him for it, David Spade had the perfect reply: "Because comedians don't have a medical chart for everyone in the audience."

The prevailing point of the *USA Today* article is that, while we have been doing a pretty good job of shutting down speech that is egregiously offensive, people tend to be more lenient when it comes to jokes. To which I say: No shit. People *are* more lenient with jokes than they are with, say, hatefully employed racial slurs, but isn't that good? How could it not be? It is treated differently because it *is* different. That ex-boyfriend calling me a "rapidly aging idiot" *is* different from Chris Rock's joke, or from the joke Greg Gutfeld made when he told me I looked like Tanner from *Bad News Bears* without my extensions in. I didn't Jada-like roll my eyes when Greg said that, but I *was* angry—only because I hadn't thought of it myself first. Nightmare of a Boyfriend's intention was to hurt me and bring me down so he could more easily control me, while Rock's and Gutfeld's intention was laughter. In Rock's case, in fact, laughter wasn't just his intention, it also was his *job*. (It's also worth noting that Nightmare's abuse had, on occasion, gone beyond words.)

It's ridiculous for Hamby to talk about "words as violence" while also championing the advantages of communication—because you simply cannot believe both at the same time.

The very idea that words are violence inherently shuts down communication. If something *is* violence, then responding to it with violence isn't violence at all. It's self-defense. Right? I'm not a general or a State Department head, but isn't that how foreign policy decisions are supposed to be made? Diplomacy is considered appropriate when a conflict is limited to an exchange of words, because an appropriate response to words is more words. Once a violent act occurs, however, then that is when violence is considered acceptable in return.

You can't say at one moment that words are violence, only to say in the next moment that communication is king, because the two are diametrically opposed. Make no mistake: The entire goal of the "words are violence" crowd is not to encourage communication, healing, or understanding; it's to wage a sort of tyranny in which nothing is off the table in stopping speech that makes them feel uncomfortable. Once you say that words are violence, the discussion almost always ends.

In fact, many of the "words are violence" crowd are quite honest about this. In March 2022, police were actually forced to protect a conservative speaker at a Yale Law School event. The event, ironically, was supposed to be demonstrating the value of being able to work with people with whom you may disagree. The school's Federalist Society had featured Kristen Waggoner, general counsel at the Alliance Defending Freedom, which advocates for socially conservative/religious causes, and Monica Miller, representing the American Humanist Association, which "advocates progressive values and equality for humanists, atheists, freethinkers, and the non-religious across the country." The pair had worked together on a case for a plaintiff suing for his right to profess his Christian faith on his campus and won—again, aiming to make a point for all of these future lawyers about working together with those with whom you may not agree on everything to achieve a common goal.

They weren't having it. Protesters shut down the event, shouting, "I'll fight you, bitch!" and blocking the only exit. Two Federalist Society members even said the protesters were grabbing at them as they were trying to leave. Police had to come escort the panelists out.

Later, a whopping four hundred law students signed a letter saying that having the police there amounted to putting the very lives of the school's gay community in danger because "LGBTQ people are six times more likely to be stopped by the police." The fact that police were there because others feared for their physical safety did not, apparently, matter.

The response of the protesters—getting physical and threatening to engage physically further—was absolutely the result of the "words are violence" mentality. The Christian conservative worked for an organization that advocated against things like gay marriage. Therefore, her presence on the campus was violence.

The event certainly did prove a point, just not one that the organizers had intended. It had intended to prove that people with opposing viewpoints can still work together, but instead it proved that the "words are violence" crowd makes collaboration, or even communication, completely impossible.

In the wake of The Slap, some comedians have expressed concern that they, too, might get punched in the face onstage for making an errant joke. What's more, many of the people who made these premonitions also saw them as justified about a month later, after a knife-wielding man charged at Dave Chappelle during his set at the Netflix Is a Joke Festival—especially after Knife Dude admitted that his motivation for the attack had, indeed, been that he found some of Chappelle's material "triggering" to him as a bisexual man.

The physical safety of comedians is always a legitimate concern, but the truth is, by "always," I mean that this was true even before The Slap. For example, less than a week before the Oscars, a man rushed the stage at Win-River Resort & Casino in Redding, California, while a gay black comedian named Sampson McCormick was performing and punched him in the face—saying, according to McCormick, "I'm going to beat your Black ass." McCormick claims that he then, in turn, repeatedly hit his attacker and slammed him into a table before security got involved.

Honestly, when it comes to The Slap, I see discussions about the future physical safety of stand-up comedians during their sets as just the tip of the iceberg in terms of what we should talk about. The main issue is that the incident, and especially the defense of Smith, is a symptom of a much larger problem.

Think about it: In many ways, Rock kind of got off at least comparatively easy. That's not to minimize what he went through, because it was awful. He was assaulted onstage in front of millions, had to watch the guy who assaulted him receive a standing ovation shortly thereafter, and then hear that a majority of people felt *he* was more wrong.

Yeah, Smith eventually apologized. And sure, Rock did have some pretty great support. Even though the majority may have said that Rock was more wrong, many others spoke out in his defense, and he even sold more tickets for his tour in the wake of The Slap than he had the week before. He definitely deserves that—especially considering the way he was able to keep his composure. I know from personal experience I could not have done that. Years ago, I was at the Union Pool bar in Brooklyn, waiting my turn to speak at my friend Ben Kissel's Brooklyn borough president campaign event, when a man walked in and tapped me on the shoulder. I turned—and then, without a single word, he dumped a giant bottle of water all over my head. I stood there as he then whipped the rest of it into my face and my eyes before walking out. I was so stunned and so upset, I couldn't stop sobbing long enough to speak at all.

But getting slapped in the face is far from the worst thing that can happen to a comedian who makes an errant joke these days. You can lose work. Gilbert Gottfried, Norm Macdonald, Shane Gillis—it happens, and even if it doesn't, you can expect your character and worth to be dissected, questioned, and trashed, just as part of your job. There was one thing that Hamby was right about: Rock "probably can't feel that slap anymore." For others, it can be much worse.

Hell, accepting the view that words are violence has been used to literally justify murder in the past.

A recent example: *Charlie Hebdo*. In 2015, seventeen people in the offices of the French satirical magazine were murdered in a series of terrorist attacks because the publication had made jokes about the prophet Muhammad.

Is it the same? No, of fucking course not. Although I'm sure that some blogger somewhere will write a think piece blasting what I just said without clarifying what I meant by saying it. Bitch-slapping someone, or defending someone who bitch-slapped someone, is not the same as murdering more than a dozen people. But here's what should scare you: The logic actually kind of is. A slap isn't the same as a murder, but if you truly believe words are violence, then violence is an appropriate response to words.

To the terrorists, the attack was an appropriate response. To the terrorists, the satire was violence. So, the terrorists did what they did not only to punish people, but also to ensure that other people would not also make the jokes that they, *Charlie Hebdo*, did.

It may seem like a uniquely extreme example, but humans have actually treated words as violence for most of our history. From the caveman days all the way through the Civil War, dueling to the death was a socially acceptable way to deal with a dispute. If you consider words violence, you're not a forward-thinking progressive; you're a knuckle-dragging troglodyte.

It's only as we have become more modern and civilized over the past few hundred years that we have moved away from this, opting to instead respond to words that insult us with words. Use your voice, not your fists, to say that something hurt your feelings. Considering words, even *jokes*, to be potential "violence" makes the stakes of bothering to communicate too high to be worth it. If a joke can be "violence"—and, therefore, justifiably responded to with "violence"—then why bother? Plus, if it can be "violence" even by accident, like in the sense of a misfired joke, then how absolutely terrifying is that?

The words-are-violence crowd doesn't want conversation—at least not one that is an equal playing field. If your words are violence, then in any conversation in the wake of your "violent words," they are ultimately coming from a power position where their response to your violence being just words is actually them doing a favor for you. They want that. They want to make you afraid. They claim to want to be understood, but that isn't what they want.

Words are violence when they disagree, and then silence is also violence when they want you to speak up, but if you speak "incorrectly," then your words are violence there, too.

In the wake of The Slap, there were several pieces explaining that any criticism of Smith was, as one piece quoting more experts, this time experts on race, put it, "rooted in anti-Blackness"—essentially unequivocally delegitimizing any opinion from a majority of the country as not just useless, but *racist*. A piece in *Teen Vogue* by a pop culture writer by the name of Stitch, titled "The Will Smith & Chris Rock Slap Situation Is Not About You," made exactly that point before concluding with a senseless statement about the violence of words:

> *Black women are rarely protected; instead, they are the punching bags for a world that seems dedicated to reminding them that they aren't fair maidens. Sure, violence isn't the best response to misogynoir . . . but let's not ignore that misogynoir is violence. Let's talk about that.*

Again, throughout the piece, the author literally *demands* silence from a majority of the country's population based on skin color, and does so *outright*. It manages to conclude by presenting the same two diametrically opposed positions that Hamby did: Violence is not the best response to misogyny, Rock's words were misogyny, and misogyny *is* violence. The conclusion is senseless: If misogyny *is* violence, wouldn't an appropriate response to violence *be* violence? That's how it works when it comes to everything from what the world considers justifiable war to what the legal system considers justifiable self-defense.

The bottom line: When you say that words are violence, you inherently *are* saying that violence is an acceptable response to words, because violence is universally considered an acceptable response to violence. This places the person who deemed the words in question to be violence in an undeniable position of power: Since they've called the words violence, they can respond as such. They aren't simply saying that the words hurt them. They are giving themselves permission to respond to those words in any way they want. Since what *you* have said is violence, all options are on the table. Claiming "words are violence" is a tool to dictate and control, all while engaging in a massive fraud that they are on the side of compassion.

CHAPTER 9

# SAFE SPACES AREN'T REAL (AND THAT'S GREAT!)

People love to talk about "safe spaces" these days—whether it's someone on the Left demanding one, or someone on the Right telling someone on the Left to go cry in one. "*Go hide in your safe space, snowflake! If you can't handle it in the real world, go home and hide out under a blanket!*" As if nothing bad can happen to you while you're at home under a blanket. Being home alone is actually the most likely time for all of your friends to be hanging out without you!

A few years ago, the demand for "safe spaces" even started to infiltrate comedy, especially on college campuses.

In 2018, for example, I wrote a piece for *National Review* about a student club at the University of London reportedly requiring comedians to sign a "safe space contract" before performing at an upcoming charity event.

The contract stated:

> *This contract has been written to ensure an environment where joy, love and and [sic] acceptance is reciprocated by all. By signing this contract, you are agreeing to our no tolerance policy with regards to racism, sexism, classism, ageism, ableism, homophobia, biphobia, transphobia, xenophobia, Islamophobia or anti-religion or anti-atheism.*

The contract did clarify that the rule "does not mean that these topics can not be discussed," but that "it must be done in a respectful and non-abusive way." This allowance doesn't make much of a difference in my eyes, which I'll explain more later.

The contract made the news after one of the scheduled comedians, Konstantin Kisin, tweeted about it, sharing his absolute disgust at the idea of the doc, saying that its very title—"Behavioural Agreement Form"—made him "want to puke." And actually not because "behavioral" was spelled the British way.

"Comedy isn't about being 'kind' and 'respectful' and the only people who get to decide what comedians talk about on stage are . . . comedians," Kisin said in an interview with PJ Media. "Comedy is supposed to push boundaries, and challenge people, and comedians should be free to mock religion, atheism, and a whole load of other things."

Plus, I'd bet that the vast, vast majority of comedians—even the low-level, aspiring ones—get into the business because they want to make people laugh. (That, and they have whatever incurable personality defect it is that makes someone feel the need to get on a stage and talk.) Laughter is their intention, and they realize that intention matters. What I'm saying is, I highly doubt that comedy is just overrun, or even moderately populated, with people who are thinking: *Man, I hate disabled people. I want to pretend that I'm trying to do comedy so I can tour around shitting on them to make them feel bad!* People can get so caught up in the weeds of discussing whether or not a comedian was "punching down" with a joke that they forget to consider that the comedian might not have been intending to punch anyone at all.

Anyway, I know what you're thinking: *Who cares about England? We won the war!* Well, unfortunately, the practice of colleges designating campus comedy shows as "safe spaces" is not uniquely British. That same year, Vice did a piece on colleges in the United States taking the same approach. Throughout it, bookers and student activity coordinators at American colleges discussed, for example, how they would never, ever book a comedian who made a joke about sexual assault—even if it were a female comedian making a joke about her *own experience*—and how strict the process is to ensure that the show is a completely safe space where none of the students will get hurt. The hurt of the female rape victim who may have been using humor to work through her trauma using comedy is, apparently, inconsequential in comparison to the hurt of people who might have to hear about it.

The justification for all of this? It's what the students want, and they'd likely complain otherwise.

Kat Michael, the booker at "women-focused" Simmons University in Boston, explained it this way:

> *The kids that I am programming for are not cis het dudes, and we intentionally program towards that, because it's what the students want, and when I'm working on a contract with a—especially with a comedian—I'm very up front in saying, you know, transphobia language isn't going to be tolerated, if you say something in your set, like, we reserve the right in our contract, like, to have a conversation with you about payment, and I will also pull the microphone. . . .*

Damn, that totally makes sense. I mean, saddling these students with lifelong debt to attend these schools would be one thing, but giving them the option to risk hearing a joke that they may not like? That could *really* impact their well-being.

The inconsistent logic doesn't stop there. During the interview, Emerson College booker Jason Meier agreed that college comedy shows need to be a safe space, saying, "What a comedian does on our campus isn't the same as what they would do at an open mic at the bar a few hours later," because "a student would respond." This seems quite at odds with a piece that ran in *Expression* (Emerson's official Student Communications and Marketing publication, in case you don't know it and love it and read it every day) in 2021, lauding the school as the "the Epicenter of American Comedy" because comedians like Bill Burr, Iliza Shlesinger, and Jay Leno had all graduated from there—completely ignoring, apparently, that the college would never permit acts like theirs on Emerson's stage today. If you don't know what I'm talking about, google "Bill Burr interracial sex," or "Iliza Shlesinger black woman impression," or "Jay Leno Asians."

As all of this stuff was coming out, I had been writing columns about college students and administrators declaring seemingly innocuous things "offensive" almost every single day for years. By that point, for example, several schools had launched campaigns warning against the usage of the phrase "you guys" to describe a mixed-gender group, on the grounds that it generalizes all people as male—effectively erasing women. In 2016, a professor at Brooklyn College, part of the City University of New York, was forced to change his syllabus after a portion stating that effort was 10 percent of the grade was deemed "sexual harassment," even though anyone but an insane person would realize that "effort" probably referred to things like class participation and not to things like a hand job. If anyone were to look at that and think, *Omg! He's saying that I have to do sex stuff to improve my grade!* then that person would pretty obviously be the one with the problem, no?

But he had to change the syllabus anyway. In 2015, I even wrote about a Huffington Post column by a college student that claimed the word "too" was sexist and hurts women—because it's so often used to say that women are, for example, "too fat" or "too skinny" or "too promiscuous."

"I never realized how deeply a three-letter adverb could cut," Cameron Schaeffer wrote. (Damn shame for Cameron that adverb outrage never got the kind of heat that pronoun outrage has been getting, you know?)

Given all of this, I'm sure you can understand why the "respectful and non-abusive" exception in the London safe space contract didn't make me feel better. It's the same reason why, when Kat Michael clarified that it "would have to be a pretty fucking terrible joke" for her to actually cut a comic's mic and talk about revoking payment, I didn't feel better about that arrangement, either. What constitutes a "terrible joke" versus a "respectful" one is completely subjective, and years of researching and writing about this stuff showed me just how shockingly low the standard for "terrible" could be. Worse, such as with that professor who had to change his syllabus, it's all too often true that even the most unhinged get their way just because they said they were offended. Who knows, a comic could get onstage in a room full of the people from the previous paragraph and say, "Hey, guys! I know it's a little too hot in here, but let's all put in some effort to have a good time!" only to be dragged off the stage and subjected to a harassment complaint. "*He is seriously so fucking abusive. He literally hates women, and demanded that we all get naked after misgendering half of us!*"

Maybe that is a bit far—although, I hate to say that, because every time I say that, I end up being proven wrong. The worst predictions you can make are about how little sense nonsensical people are capable of making.

Worse, this idiotic thinking hasn't gone away—nor has it stayed confined to college campuses. In July 2022, a theater in Minneapolis canceled Dave Chappelle's sold-out show, apparently to appease the 125 people who had signed an online petition calling him transphobic.

Just hours before the show was scheduled to begin, First Avenue released a statement calling it off and apologizing for having invited Chappelle in the first place:

"The First Avenue venue team and you have worked hard to make our venues the safest spaces in the country, and we will continue with that mission," it added.

The truth is, there's really no such thing as a "safe space," because everyone is going to have a different idea of what that means. You can plan for everyone to only attack "acceptable" targets, but you'll never really know what targets are and are not "acceptable" for each person in attendance. We're all unique individuals with unique sets of life experiences, preferences, and beliefs—which means that we're all going to view any given situation in our own unique way.

In 2015, a Harvard University student inadvertently illustrated this pretty well when she wrote an op-ed complaining that her school's safe spaces weren't safe enough. In the piece, Madison E. Johnson detailed her experience attending one, which she said was a great time at first—complete with "massage circles," "deep conversations," and designated "processing and journaling" times.

For me, that's already my idea of Hell. Like, I can feel my cuticles bleeding just thinking about how much I'd be anxiety-picking at them throughout that kind of hang.

But Johnson says her "safe space"—which she defined as "one in which I feel that I can express all aspects of my identity without feeling that any one of those aspects will get me (including, but not limited to) judged, fired, marginalized, attacked, or killed"—eventually shattered. She listed a few incidents: a student in the "safe space" asking if she was "a full black"; a white poet going onstage and saying "the n-word a few times."

She eventually said she realized that "'safe space' might mean different things for different people," before ultimately concluding: "I don't know what we can do to change it, but openly acknowledging that some of the safe spaces at nice, progressive Harvard aren't all that safe for some of us sounds like a good start."

Johnson was right about the first part: The school's "safe space" wasn't really safe at all. What she got wrong, though, was her idea that the school could maybe create one if it just started trying harder. The real issue is that it's impossible.

Don't get me wrong: What would possess any white person to get on a stage and just start spouting off the n-word is absolutely beyond me. Like, that's a total asshole, right? But, as I wrote about the issue at the time: "One student might consider a 'safe space' one where she can use the n-word freely without fearing judgment, and another might consider it a place where she can be certain that she won't have to hear it." No common space can guarantee safety for every person's feelings, because you can't guarantee that every person's feelings will be the same. Actually, you can guarantee that they won't be.

Also, sometimes, some of the most safe-space obsessed are the biggest assholes out there. They call other people out as a way to shield the public from their own closet full of skeletons.

By 2019, a Canadian transgender-nonbinary comedian named Chanty Marostica, who uses they/them pronouns, had racked up an impressive stack of accolades: the first trans person to headline Toronto's Just for Laughs 42 festival in 2019, winning Sirius XM's Top Comic contest in 2018, winning Canadian Comedy Awards' Best Album in 2019, and a nomination for a Juno, which is like a

Canadian Grammy—and, perhaps most prestigious of all, winning the position of a media darling for their undying devotion to safe spaces.

Chanty was *woke*, and (*Snap! Snap! Snap!*) not afraid to show it. In the wake of their meteoric rise, they told Vice in an interview:

> *And I could see, there's a lot of cis heterosexual white men that just get really mad when you talk about men being shitty, and it's like, if you're not part of the conversation to change why men are shitty then go.*
>
> *Any time a woman has anything to say about how they feel unsafe, it's just gaslit to the point where they delete their comments. But comedy is very unsafe for women and I can say that out loud, because I look like this. And when women say that, they get disrespected or not booked. And I get to talk about it because I don't take for granted that I haven't looked like a woman for a very long time. I think that, people when they hear something about what they do wrong, their immediate reaction is to be like "No, I didn't, fuck you, you're lying!" but atonement and accountability is the only thing we can do to change. It's so hard to face your own phobias and react in a way that's conducive to change, but that's the only way we can grow as people, is unlearning all the garbage we've been told our whole lives.*
>
> *I was the only woman, and then I was the only queer person, and then I was the only trans person, I've always been the only one, and it's so jarring to be so alone, and to feel unsafe.*

But Chanty Marostica—who had spent their career alleging the worst of others—was about to *lose* their career due to others' allegations about *them*.

A piece in Quillette shares a story about how, in 2019, Marostica heard another comic, Matt Billon, tell a joke onstage that Marostica felt was transphobic during a show that both of them were on. I wasn't there, but my sources (the Internet) say Billon's joke was something about how men being in women's sports would make those sports more interesting to watch or something. Marostica heard the joke, wrote their transphobia allegation on a napkin in the greenroom for Billon (and everyone) to see, and then promptly left the show.

Billon was, of course, totally fucked. He had upset the scene's Brave and Inspiring Hero, and no amount of apologies or attempts to make it right were any match for Marostica's, uh, *compassion*. He was smeared as a transphobe, which affected him in terms of Facebook posts saying so, and people denying him opportunities without saying so. No one had his back, presumably out of fear that they'd be dragged down with him. Amid Billon's cancellation, people continued to rally around Marostica—at least until later in 2019, when Marostica made a post calling a Canadian comedy club's decision to book "abuser" Louis C.K. a slap in the face to women everywhere and "unsurprising, lazy, and archaic."

It wasn't the post itself that caused problems for Marostica. I mean, the post was classic Chanty! Rather, it was the comments on the post accusing Marostica of sexual abuse.

A comedy club employee accused them of abusing several people. A female comic accused them of being a "predator and a gatekeeper," claiming she herself was one of Marostica's victims. It's unclear if any of these online accusations turned into real-world claims, but the Internet took them very seriously.

Eventually Marostica issued a buzzword-salad apology and essentially faded from public discourse. Matt Billon died by suicide in November 2021.

It seems possible that Marostica was a predator parading around as a social justice champion. It's not the first time someone hasn't walked the talk. In 2019, New York governor Andrew Cuomo signed multiple pieces of legislation that "strengthened protections against discrimination and harassment," only to resign over his own sexual harassment scandal in 2021. Chrissy Teigen was once widely celebrated as a woke, progressive darling, largely for her sassy, Twitter-clapback opposition to Mean Bully President Donald Trump and anyone who supported him—even demanding a boycott (some might say, a cancellation) of Equinox and Soul Cycle in response to reports that the companies' developer planned to host a fundraiser for then-President Trump's reelection campaign. All of this, of course, for Teigen to *still* be whining a year later about how devastating her *own* cancellation was after she got busted DMing a minor—Courtney Stodden, who uses they/them pronouns—and telling them to kill themselves, which sounds suspiciously like something a bully might say. (In Teigen's defense, she's so quirky! A model, but quirky!)

I'm not saying that everyone who champions the "safe space" mentality is a predator, or even a jerk. Certainly, at least some of these people must have good intentions. Still, it would kind of make sense that so-called safe spaces would be a perfect place for predators to hang out. No space is safe, but being in one that bills itself that way just might make you believe it. It can easily give you high expectations for a false sense of security, prompting you to let your guard down.

What's more, many of the people who levy charges of racism, transphobia, homophobia, or sexism know that they can gain control of a situation by doing so. After all, sometimes nothing puts you in a position of power over another person quite like calling him a white supremacist or a transphobe. If he argues with you, then you can use his arguing as evidence that you're right, and any people who don't have your back know that they're risking the same fate as the person you just took down—especially if, like Chanty, you can also claim one or more victim classes for yourself.

It's best to admit that a safe space is never possible. It's ridiculous to demand one, and especially so when it comes to comedy.

Besides the obvious, calls for "safe spaces" often go hand in hand with calls for the usage of "trigger warnings," alerting about potentially disturbing content—purportedly to be sensitive, helpful, and kind to the people who have gone through related traumas.

First, one thing about trigger warnings in general: They don't work, and might even be harmful. (Yeah, bro, I don't use the word "purportedly" for no reason!) Harvard psychologists Payton Jones, Richard McNally, and Benjamin Bellet conducted a series of studies on trigger warnings and found them to be useless at best, and actually harmful to trauma survivors at worst.

The study's abstract states:

> *"We found no evidence that trigger warnings were helpful for trauma survivors, for those who self-reported a PTSD diagnosis, or for those who qualified for probable PTSD, even when survivors' trauma matched the passages' content."*
>
> *"We found substantial evidence that trigger warnings countertherapeutically reinforce survivors' view of their trauma as central to their identity."*

Trigger warnings are not helping the people they claim to help and may even be harming them. To me, that seems like a good enough reason to stop using them. Like, don't you think?

Regardless, with comedy in particular, using them would obviously ruin everything. The element of surprise, after all, is necessary for a good joke. If you are expecting a punch line, then it won't work as a punch line at all. (Or, as Norm Macdonald put it:

"Comedy is surprises, so if you're intending to make somebody laugh and they don't laugh, that's funny.")

Think, for example, about *Borat* (yeah, I know, I bet you haven't since 2007), when Sasha Baron Cohen's character says, "In Kazakhstan the favorite hobbies are disco dancing, archery, rape, and table tennis."

Would this joke have worked if, instead, the screen had gone black in the middle of the movie, and the words "TRIGGER WARNING: SEXUAL VIOLENCE" appeared, and *then* Cohen said, "In Kazakhstan, the favorite hobbies are disco dancing, archery, rape, and table tennis"?

Obviously not, right? Thank God Cohen decided to *not* do that, even though he was risking his ability to be invited to do college gigs a decade later when standards would have changed in a way he could have never understood at the time.

I mean, not only was the joke hilarious, but it was also a great social commentary on the horrific treatment of women in countries like Kazakhstan. Part of the reason that it works is the element of surprise, using shock to introduce the shockingly horrific, all while using humor to make it palatable enough to digest. But these days, a joke like it could never fly on a lot of college stages, not even if the person making it was a female comedian from Kazakhstan who had been raped by a Kazakh.

It's a good thing to see injustice in the world and talk about it. It's not bad to have hurt feelings and to share them. If you see injustice, or if your feelings are hurt, then yeah, you should absolutely say something. "The more speech the better!" is true, and that includes speech about feelings.

What's unhealthy and destructive, though, is to demand that other people bend to your feelings just because you have them. What's worse is how people have somehow managed to brand this practice as "compassion." Demanding that the entire world and its entire discourse revolve around your feelings isn't altruistic. It's selfish. And demanding that people who don't see the world how you do be ruthlessly shamed without forgiveness isn't heroic. It's bullying.

It's kind of why I have always hated the whole "snowflake" thing.

It can be too mean, for example, when it's used to ridicule someone just for having feelings, or for being upset about something. Caring about children stranded alone at the border, for example, doesn't make you a "snowflake." It means that you care about children stranded alone at the border, because you realize how unbelievably traumatic that must be for them. I remember someone calling me a "snowflake" on Twitter once because I was upset about my mom being dead. Like, okay? Checkmate, I guess?

It's okay to have feelings; it's normal to have feelings. If you stand for "free speech" and "more speech," then you should know that speech about feelings is all part of that. I know that a common refrain is "Facts don't care about your feelings," and that's completely true. But it's also true that, in some cases, feelings don't care about your facts. (Actually, it perfectly describes how I handled every relationship that I had in my twenties!) Sure, it's better to be rational, but it's also pretty impossible to get there without being able to talk about it. If you're feeling something, say it. That doesn't make you a "snowflake." It makes you human.

Then there's the other way that people use "snowflake": to disparage people who want to shut down speech. I hate that, too, because those people deserve to be called something way worse. They're not snowflakes—they're oppressive, and they're assholes, and they're selfish, and they're bullies. It's always so weird to me when people are shocked by something like what happened to Chrissy Teigen. "*How could this social justice warrior have been a bully all along?!*" To me, it's *not* shocking. It didn't take those Courtney

Stodden DMs to reveal Teigen as a bully. She had already revealed herself as one with the way she called for the cancellation of entire corporations (thereby jeopardizing the livelihoods of all their employees) simply for having a developer with whom she disagreed politically. Who would do something like that *except for* a bully? That's not compassion. Compassion is approaching differences with understanding and grace, not using them to wield power over your enemies.

Again, I'm not talking about people who want people to treat others with respect, or who want to see progress on issues like racism and sexism. I'm talking about people who use words like "racism" and "sexism" to place themselves in a position of authority over others.

It's extremely important to push back on this sort of thing, which is why I was so glad to see that, in May 2022, Netflix released a memo titled "Netflix Culture—Seeking Excellence," which stated:

> *Entertaining the world is an amazing opportunity and also a challenge because viewers have very different tastes and points of view. So we offer a wide variety of TV shows and movies, some of which can be provocative. To help members make informed choices about what to watch, we offer ratings, content warnings and easy to use parental controls.*
>
> *Not everyone will like—or agree with—everything on our service. While every title is different, we approach them based on the same set of principles: we support the artistic expression of the creators we choose to work with; we program for a diversity of audiences and tastes; and we let viewers decide what's appropriate for them, versus having Netflix censor specific artists or voices.*
>
> *As employees we support the principle that Netflix offers a diversity of stories, even if we find some titles counter to our own*

> *personal values. Depending on your role, you may need to work on titles you perceive to be harmful. If you'd find it hard to support our content breadth, Netflix may not be the best place for you.*

I thought this was awesome—especially considering how, for example, hundreds of Netflix employees had staged a walkout in the fall of 2021 over the fact that the streaming service had allowed a (yeah, this guy again) Dave Chappelle stand-up special featuring jokes about trans people to remain on the platform.

The only thing that bothered me about some of the dissent to the walkout was the characterization of them as a bunch of pussies, with Chappelle himself referring to the upset members of the LGBTQ community as being "too sensitive." After all, there's nothing sensitive about demanding that the world afford you special treatment simply because of your identity. They weren't just expressing their feelings; they were attempting to exercise collective power to control the expression of others. As Ricky Gervais put it in his own Netflix special, which also prompted backlash for its jokes about trans people, "I talk about AIDS, famine, cancer, the Holocaust, rape, pedophilia. But no, the one thing you mustn't joke about is identity politics. The one thing you should never joke about is the trans issue. 'They just want to be treated equally.' I agree. That's why I include them."

There's a huge difference between asking to be treated equally and demanding to be uniquely untouchable. Actually, they're diametrically opposed! It's one thing to be sensitive, and quite another to be so far up your own ass that you consider any joke at your expense to be an unforgivable sin. If you view the world that way, you're not a sensitive snowflake; you're a self-obsessed, petulant tyrant.

Plus, although it may seem at odds with our current discourse: If you really want to be benevolent, then you should not only recognize the fact that a "safe space" doesn't exist, but you should also preach about the value of exposing yourself to uncomfortable ones. Think, for example, of that piece in the *Atlantic* that I cited in my chapter on the Slap—explaining how hearing tough things can actually make people stronger, giving them coping mechanisms to help them more easily get through tough situations in the future.

But it's more than that. Although it may seem counterintuitive, being in an uncomfortable situation can actually make you *happy*. Research covered in an article in *Forbes*, aptly titled "Why Feeling Uncomfortable Is the Key to Success," found that "[p]utting yourself in new and unfamiliar situations triggers a unique part of the brain that releases dopamine, nature's make-you-happy chemical. Here's the mind-blower; that unique region of the brain is only activated when you see or experience completely new things."

Got that? It doesn't just make you feel good. It makes you feel good in a way that is *only* possible when you're experiencing something new, as uncomfortable as it may be.

"While it may not feel like it in the moment, a little bit of discomfort goes a long way in terms of personal development," *Forbes* explains. "Sure, no one likes feeling uncomfortable, but it's a big part of improving your performance, creativity, and learning in the long run."

Considering this, encouraging others to demand so-called safe spaces ultimately amounts to encouraging them to hinder their own growth and happiness.

To me, it makes sense. When I look back at the most transformative times of my life, the moments that I think of are the ones where I was the least comfortable, many of which I talk about in this book—the shitbag, the Coney Island Breakup, the time I spent struggling in Los Angeles. Honestly, performing stand-up comedy, regardless of how many times I did it, pretty much always terrified me—and I'm also not sure where I'd be if I'd never done it.

The truth is, the fact that comedy is a particularly unsafe space isn't a bug that needs to be fixed; it's a virtue of the art. Hearing jokes about sensitive topics can be uncomfortable, because it flies in the face of the sacredness with which we are "supposed" to approach those subjects, but that isn't necessarily a bad thing so much as it is an opportunity to grow.

CHAPTER 10

# ON APOLOGIES AND APOLOGIZING

In 2015, I made some little jokes about *Star Wars* and *Star Wars* fans on *Red Eye*. All I said was: "Yesterday I tweeted something, and all I said was that I wasn't familiar with *Star Wars* because I've been too busy liking cool things and being attractive" and "I have never had any interest in watching space nerds poke each other with their little space nerd sticks, and I'm not going to start now."

Then I went home, and kind of forgot about it. I mean, sure, I got a few death threats after I posted that tweet that I referenced during the segment, but everything was mostly fine—until about a month after that. See, a *Star Wars* supernerd found my *Red Eye* comments and made a more than ten-minute-long video berating me for my jokes, and even did his own supernerd version of fact-checking me by posting some weird slide show of hot chicks wearing *Star Wars* T-shirts.

After that video, I was absolutely inundated. There were lots of death wishes—I remember someone telling me that he hoped I would die from the same disease that had just killed my mother a year earlier—and also straight-up threats, such as "Tomorrow at 8 am, you're dead. Shouldn't have said what you said. Call the FBI, call the cops, call whatever you want, nothing can stop the onslaught that you are about to face." (Yeah, that's an actual email that I got. Under "Name" it said "Grim Reaper" and under "Subject Line" it said "Death.")

There were rape threats, too—several of which involved a lightsaber and my asshole, proving that these guys really *were* such nerds that even their threats of sexual violence were nerdy. The story was so huge that it was a trending topic on Facebook, and I even heard from a friend that there was a Buffalo Wild Wings trivia question about it.

The major honor of being a Buffalo Wild Wings trivia question aside, a lot of people in my life were afraid for me, and they encouraged me to apologize so that the whole thing could blow over.

Instead I wrote a piece for *National Review* titled, "I Will Not Apologize for Making a Joke About *Star Wars*."

In the piece, I explain:

> *A lot of people are clearly a lot of upset. But guess what? I'm not apologizing. Why? Because the all-too-common knee-jerk reaction of apologizing for harmless jokes after overblown hysteria is ruining our culture. This political-correctness obsession threatens free speech, and I absolutely refuse to be a part of it.*
>
> *Bottom line: If you are telling me that I should die and/or apologize for making a joke about a movie you like, then you are too sensitive. You have the problem, not me.*
>
> *I'm sick of oversensitive mobs in our overly sensitive society bullying people into saying that they're sorry over jokes—even if the*

> *subject of the joke is something as serious as* Star Wars. *So, for that reason, I will not apologize.*
>
> *Also, I just don't have time. After all, I am too busy liking cool things and being attractive.*

I'm glad I handled all of it the exact way that I did. I wasn't sorry, so I didn't say I was, and the whole thing sort of died down on its own anyway. I say "sort of," by the way, because I did actually get an email in 2020—a full five years after the joke—saying (all errors his; yeah, "his"; I'm assuming it's a man):

> You are an incredibly oppressive bitch for the comments you made about star wars fans. You are not qualified to make an assumption about star wars and it being "Nerdy" if you have not seen it. You are so fucking stupid and i am going to request that you be fired for being oppressive an offensive. You are a fucking cunt and i hope you get herpes. Fuck yourself you fucking autistic retard. Oh and your marriage is a sham.

(I know. It's the 2020s. Are we *really* still stigmatizing herpes?!)

There's been a lot of discussion about how the refusal to *accept* apologies can hinder speech, but the truth is, an obsession with *making* them can do that, too.

When we routinely apologize for things that we don't actually believe are wrong, we set moral standards for our culture that don't line up with what we actually believe. Every time we apologize for something that we don't really think is wrong, especially publicly, we're just adding another nail to the plank that someone is going to be forced to walk off of someday. If we condition our culture to believe that any joke that upsets any person demands a groveling, on-the-floor apology, then we are going to get exactly what we asked for—fewer jokes, less honesty, and a really unforgivingly low fuckup threshold for ourselves.

Also, if what you're looking for from an apology is for people to actually forgive you, then you might be extremely disappointed.

In February 2021, Chris Harrison, the longtime host of *The Bachelor*—a show where people try to find true love by dating a bunch of people in front of each other, banging a few, and then proposing to one after just having banged those others—got himself into big trouble for cautioning against canceling one of the front-runners on the season featuring the franchise's first black Bachelor, Matt James.

As the season was airing (which was months after it had been filmed), photos surfaced showing the eventual winner, Rachael Kirkconnell, attending an Old South antebellum party in 2018. To be clear, I personally believe that dressing up in a slave-owner outfit to go get drunk on an old slave plantation is an extremely fucked-up way to party. I don't own a hoop skirt, nor do I know where I'd buy one. Probably on Amazon.

Rachel Lindsay, an *Extra* correspondent who was also the first black Bachelorette, asked Harrison about the photos on a podcast before Kirkconnell herself had weighed in on them.

"I saw a picture of her at a sorority party five years ago, and that's it. Like, boom," Harrison said. Unsure if he had really hammered his point home, he added: "I'm like, 'Really?'" (I bet it's impossible to work in reality TV without winding up talking like that. I have enough issues just because of how much I watch it.)

When Lindsay replied that it "wasn't a good look," Harrison contested: "Well, Rachel, is it a good look in 2018? Or, is it not a good look in 2021? Because there's a big difference."

Lindsay insisted that it was "not a good look ever," adding, "If I went to that party, what would I represent at that party?"

Harrison replied:

> *You're 100% right in 2021. That was not the case in 2018. And again, I'm not defending Rachael. I just know that, I don't know, 50 million people did that in 2018. That was a type of party that a lot of people went to. And again, I'm not defending it. I didn't go to it.*

Controversy, of course, ensued, and Harrison apologized for his comments:

> *"To my Bachelor Nation family—I will always own a mistake when I make one, so I am here to extend a sincere apology," the 49-year-old host began on Wednesday, February 10, via Instagram. "I have this incredible platform to speak about love, and yesterday, I took a stance on topics about which I should have been better informed. While I do not speak for Rachael Kirkconnell, my intentions were simply to ask for grace in offering her an opportunity to speak on her own behalf."*

That first apology was then followed up with another, with Harrison saying he had "spent the last few days listening to the pain" caused by his words, that he was "deeply remorseful" for how his "ignorance did damage to friends, colleagues, and strangers alike," that he feels his failure with "every fiber of [his] being," and that he would be "stepping aside" from the role as host "for a period of time." During that period, in March, he appeared on *Good Morning America*, where he apologized yet again, saying he "made a mistake," apologizing directly to Lindsay and "the black community."

"Antebellum parties are not okay—past, present, future," he said. "Knowing what that represents is unacceptable," adding, "I am committed to the progress—not just for myself, also for the franchise."

But his time with the franchise was already over. Despite his multiple groveling apologies, Harrison was not able to ever move past the controversy and return to work, and *The Bachelor* has another (white male) host now.

(After a break, James forgave Kirkconnell. As of this writing, the two of them are still together.)

No doubt: Seeing these sorts of controversies play out this way could be a major reason that some people might think that apologizing just isn't worth it, especially since Harrison is far from the only example. Chrissy Teigen's two public apologies (again: for, among other things, direct-messaging a then-teenage Courtney Stodden to tell them that they should kill themselves) weren't enough to get her those cookware deals back—reducing her from the megarich wife of an A-list singer to the megarich wife of an A-list singer without cookware deals. The same goes for Kathy Griffin: no apology or tearful press conference about how Trump "broke" her in the wake of her fake-decapitated Trump-head photo shoot was enough to get her CNN New Year's Eve hosting gig, tour dates, *or* Squatty Potty commercial back—which may be why she *un*-apologized a year later on *The View*, saying, "I take the apology back. Fuck him."

Trump, of course, counts himself among those aboard the Don't Apologize Train—even chastising Joe Rogan for having done so in February 2022.

If you haven't heard of Rogan, he's basically this guy who has a podcast where he smokes weed and makes $100 million more than all of the other guys who do that. (All the ones I dated always *needed* money.)

A big difference between Rogan and those other guys, of course, is that Rogan gets such big guests that people have to talk about it. I don't mean "big" in the sense of "*You've seen her on* Euphoria, *now listen as she tells the story of losing her virginity, which I will ask her about only so I can make it about me*" (that would be *Call Her Daddy*) or even "big" in the sense of Hollywood superstars, emphatically sharing the only opinions they're allowed to have if they want to keep that Hollywood-superstar status. In fact, Rogan's guests are "big" in the opposite sense: You *won't* hear from some of them anywhere else, because other platforms won't have them—except maybe as a punching bag, invited by the host to shout over as evidence of that host's brave devotion to the Correct Opinion.

Rogan approaches his interviews with nothing to prove. He has genuine, curious conversations with people, including people you're not supposed to talk to like that—such as Dr. Robert Malone, an mRNA vaccine researcher who has since become critical of mRNA technology. Many in the media worked hard to cancel Rogan over this, insisting that these conversations were a threat to public health. It didn't work, so then out came a montage of old clips of Rogan quoting the n-word and making racially insensitive jokes. Rogan apologized, saying that the clips had been "taken out of context" but that discussing the matter was still the "most regretful and shameful thing that I've ever had to talk about publicly."

Then, Donald Trump decided to release an official statement:

> *Joe Rogan is an interesting and popular guy, but he's got to stop apologizing to the Fake News and Radical Left maniacs and lunatics. How many ways can you say you're sorry? Joe, just go about what you do so well and don't let them make you look weak and frightened. That's not you and it never will be!*

I didn't really like Trump's statement—and that's not just because it wasn't as funny as the one he released calling that horse who won the Kentucky Derby a "junky" for his positive steroids test. (I don't care who you are, you have to admit that a former president releasing a formal statement to call a horse a "junky" is funny.)

It's also not because I don't understand the impulse to view apologies the way Trump does. Again, these days, it seems like whatever apology you may give—no matter how contrite or even groveling—it never ends up being enough anyway, so why bother?

A 2012 paper in the *European Journal of Social Psychology* by researchers Tyler G. Okimoto, Michael Wenzel, and Kyli Hedrick seems to suggest that Trump is not the only one who associates apologizing with weakness.

"We do find that apologies do make apologizers feel better, but the interesting thing is that refusals to apologize also make people feel better and, in fact, in some cases, it makes them feel better than an apology would have," Okimoto told NPR.

"When you refuse to apologize, it actually makes you feel more empowered," he added. "That power and control seems to translate into greater feelings of self-worth."

In an interview with the *New York Times*, Okimoto explained that "[i]n a way, apologies give power to their recipients."

"For example, apologizing to my wife admits my wrongdoing; but apologizing also gives her the power to choose whether she wants to alleviate my shame through forgiveness, or increase my shame by holding a grudge," he said.

Of course, this single piece of research shouldn't lead anyone to the conclusion that apologizing is always the wrong move. In fact, Okimoto even clarified to NPR that he himself has no problem saying sorry—also telling the *Times* that refusing to apologize could ruin "the trust on which a relationship is based," and that ultimately, "digging your heels in actually shows people your weakness of character rather than strength."

The potential benefits of an apology, after all, are well documented and widely recognized. The NPR piece about Okimoto's research also touts the "huge" "intrapersonal benefits" of apologies, including the power to "renew bonds not only between people but also between countries." The *Times* piece notes: "When you refuse to admit your mistakes, you are also less open to constructive criticism, experts said, which can help hone skills, rectify bad habits and improve yourself overall." A piece in the *Atlantic* by Noah Berlatsky points out that "[t]he reason to teach kids to apologize isn't to make the wrong-doer feel better. It's to make the person wronged feel better."

So the answer, of course, is not simply to never apologize for fear of looking weak. There has to be some kind of middle ground, which I think Sarah Silverman expressed perfectly in *Comedy Gold Minds with Kevin Hart* on Sirius XM in 2021:

> *Apologizing doesn't shame me. It doesn't scare me. It makes me feel free. I never understand how it's hard for people. I apologize when I'm sorry. I don't apologize when I'm not sorry, but I'm fucking sorry a lot because comedy isn't evergreen. And to that point, I'm not a bad person because I did a bad thing in the context of what we've learned in the world.*

It is the best way to look at it: Always apologize when you feel bad and want to make it right, but never apologize when you don't feel bad so much as feel like that is what you are supposed to do to calm pissed-off people down. It's why I did, for example, absolutely apologize after that Kimmel thing: I wanted to make it clear that that wasn't what I meant, and I felt bad about the possibility of anyone getting hurt by another interpretation.

Of course, since you've already read that chapter, you know that my apology didn't really help me any. People still went after me, especially since my besties over at Raw Story decided to totally omit the fact that I had even apologized—and, when called out on it, only edited it to say that I had "later" apologized, missing the fact that I'd done so the very next chance that I had to speak.

Because of the brutality of the response, I actually did have people ask me if I regretted apologizing, or even straight-up tell me that I should not have apologized. One of those people was even my forever-unrequited high school crush; he messaged me on Facebook! But I still didn't regret it . . . because I didn't apologize because there was a mob who wanted me to, only because I wanted to. Also, I'm not into the same guy as I was when I was fifteen.

Plus, if you think about it, neglecting to apologize when you want to apologize is really just another form of self-censorship.

Think a little more about what Trump said to Rogan in explaining why he should not have said sorry: *Don't apologize, because the mob will say you're weak.* In other words? Although he may not see it this way, Trump is really objectively telling Rogan that he should worry about what he says simply because of a mob's possible reaction. If you want to say you're sorry, but neglect to do so because you're afraid a mob will call you weak, then you're censoring your speech to bend to the will of that mob. We would live in a better, healthier, more honest society if we didn't bend our speech around angry mobs—and that includes apologies.

I can admit that there are things in the past that I wish I had apologized for, but didn't, because I'd been advised not to or didn't know how or thought it would be better for it to blow over. The thing is, though, I regret those times—and, like Okimoto, view them as incidents of weakness rather than strength.

Not that apologies are always strength. Over-apologizing, after all, can reduce the respect that you have for yourself, and even make others feel the same. As psychotherapist Beverly Engel explains in her book *The Power of an Apology*, "over-apologizing isn't so different from over-complimenting: You may think you're displaying yourself as a nice and caring person, but you're actually sending the message that you lack confidence and are ineffectual," adding that "[i]t can even give a certain kind of person permission to treat you poorly, or even abuse you."

And, like, yeah. I mean, I remember the time I took that Nightmare of a Boyfriend who didn't want people to know we were together (that's not, by the way, just my interpretation of his wants—that was the reason he gave me for why I couldn't come over to his apartment, not even after years of seeing each other or me blowing a ton of money to take us on a luxury vacation to Tulum) to a Yankees game. I posted a picture of the field from our seats, and he got furious. *He* had just posted a picture and, based on the logistics, "people" would figure out that he and I must have been there together. (The horror!) Rather than understand the objective insanity of his anger, I spent the rest of the afternoon apologizing and hating myself for being so stupid as to ruin what could have been a perfectly nice afternoon. Worse, my dumb ass would manage to bungle our beautiful love again later that summer—by asking him if I could please stay at his place when my air conditioner was broken and I couldn't carry the replacement up the stairs by myself and my apartment was above 90 degrees and climbing. I was *such* a pain in the ass, and he was *such* a gentleman, that he actually left his Important Work Night (hanging around open mics to "network") to help me put in the new unit. Even though he was cruel to me the entire time he was doing so, it was nothing compared to what a needy, unreasonable jerk I was for asking if I could spend one single night at my longtime boyfriend's apartment, even if I had promised to be quiet. I spent days apologizing for that one.

In other words? I'm thinking Engel's right. Oh and also? A toxic relationship is a hell of a drug.

There's also this: When you apologize because you Feel Like You Have To, there's a pretty good chance that your apology isn't going to come off as genuine. That's the thing about acting like a fake-ass bitch: sometimes people happen to notice that you're being a fake-ass bitch.

In a lecture to his students shortly after learning he had pancreatic cancer and not long to live, Carnegie Mellon University professor Randy Pausch said, "A bad apology is worse than no apology." The phrase became a chapter title in his bestselling book, *The Last Lecture*, discussing how truly insulting it feels to get a non-apology apology. Because we've all been there, right? Someone fucks you over ten ways till Sunday, and you're devastated and crying, and they just say, "I'm sorry you feel that way"?

People apologizing when they don't really mean it leads to so many generic, meaningless, cookie-cutter apologies that no one pays attention to any of them. You do, after all, already know what it's going to say. "I was wrong, and I am sorry for the pain that I caused, and I will do the work." It's not just that no one cares anymore. It's also that it's hard to *make* anyone care—because we're being asked to care too often.

Or, as the *New York Times* put it in a piece about this issue, titled "He's Sorry, She's Sorry, Everybody Is Sorry. Does It Matter?"

> *Social scientists have deemed this concept normative dilution—the idea that it's possible for a thing to become so normalized that we become cynical about it, even as we demand it. But that cynicism can make us less likely to forgive, in turn rendering an apology, even an authentic one, useless.*

(I think my explanation was more eloquent.)

Perhaps my favorite example of an obviously meaningless apology came from Selena Gomez after some people accused her of throwing shade at her ex-boyfriend's wife, Hailey Bieber, by posting a skin care tutorial on TikTok. Apparently skin care tutorials are *Hailey's* thing; *Hailey* posts skin care tutorials all the time, so

*Selena* posting a video of her silently spraying products on her face and lotioning it up must have been a subliminal slam at *Hailey*. Rather than just ignore it, or say something like "*You people are insane. No one has trademarked Putting On Lotion*," Gomez first disabled comments on the video and then apologized in a TikTok comment of her own, saying, "This is why I believe in taking care of your mental health. Guys no idea what I did, but I really am sorry. Zero bad intention. Deleting soon."

"*I have no idea what I did, but I'm sorry.*" Like, uh, then you're *not* sorry? If you don't even know what you did, you can't be sorry for it, and that's okay. Just don't say you're sorry when you're not, especially when your apology is literally admitting that you're not.

The truth is, it really is harder for an apology to mean anything anymore, because of how often people feel pressured to give them even when they're not sorry. That sucks for people who really do feel bad and want to make amends, because it's so much less likely that anyone is going to take them seriously. If you apologize when you're not sorry, you're not only wasting your and everyone else's time, but you're also screwing over all of the people who are going to mess up in the future . . . including you!

I know that I've talked a lot about apologies in general throughout this chapter, so I do want to make it clear that there should be a different approach when it comes to apologizing for jokes in particular. Say sorry if you're sorry and you want to, but don't forget along the way that what you're sorry for here is a joke. It's not like you banged someone's wife. Common knowledge says that you shouldn't say anything that could be construed as a defense of yourself when apologizing, and in most cases, I think that's true. When it comes to comedy, though, I think that it's okay to make it clear that your intention was a joke. (No, you can't just *say* it was a joke if what you *did* do was bang someone's wife.)

Also, a little bit of humor can sometimes go a long way when it comes to making an apology in some situations, especially a self-deprecating one. Definitely judge the situation and the other person's attitude, but sometimes nothing can ease the tension of a difficult conversation quite like it. (Again, I wouldn't do this if you banged someone's wife.)

With all that being said, I'd now like to personally apologize to:

*My roommate Matt*

*Matt's couch*

*Anyone who took the Megabus between New York and Washington, DC, from 2011 to 2014*

*The customer service team at Andy Capp's Hot Fries Corn and Potato Snacks*

*Build-A-Bear Workshop*

*Justin Verlander*

*The Ruby Tuesday at Lakeside Mall in Sterling Heights, Michigan (RIP)*

*Vinny Guadagnino*

*Every conference at the Long Beach Convention Center in 2011*

*The* Queen Mary

*Coleslaw*

*Costco*

*Brittany's dad*

*The Ann Arbor, Michigan, police department*

*Anyone who attended the Silvertide concert*
*at Harpo's in Detroit in 2007*

*Boingo Wireless*

*American Spirits*

*Anyone who has ever listened to Eminem around me*

*Aruba*

*The Duane Reade drugstore in the East Village*

*Each and every man's wife*

To you all, I am sorry—and I promise to do the work.

CHAPTER 11

# SORRY, BUT THIS ONE IS ABOUT POLITICS

One of the first times I ever watched *Red Eye*, on my then-boyfriend's brother's television, I said, "I would be great on that show."

The boyfriend's brother said, "You are a cashier at Boston Market."

(Look at me now, bitch!)

He was right, though. I was a cashier. Sure, as I mentioned earlier in this book, I was also an intern at the Fox News bureau in Los Angeles, where I would transcribe copy from interviews and tag along with reporters covering red-carpet events—snapping photos of celebrities like Ellen DeGeneres and Dr. Drew (I was huge into *Loveline* at the time) on my digital camera so I could upload them into albums on my Facebook page.

To be fair, I could see then what I was capable of, but never did I actually expect that some of the exact things I'd imagined would become a reality in just a few short years.

There were countless interns going in and out of the Fox News bureaus every summer, but I was probably the weirdest and worst-dressed one they had seen in a long time. For a quick second, imagine me as I was then: I wore cheap, floral-printed cardigans, and I'd tie rubber bands around the flared-out bottoms of my dress pants so they wouldn't get caught in anything as I was riding my bike there because I didn't have a car. (Before leaving, I'd also have to retie the rubber bands so I could then bike to that cashier shift. It was 2010; all my pants were flared.)

Ever since I was a kid, I have hoped to have a career that combined comedy and politics. I've always been passionate about politics . . . in the sense that I hate them, and what is more passionate than hatred? There is so much power in freedom, and in allowing people to make their own decisions about their own lives. Unfortunately, neither of the two major political parties running the country these days seems to agree with me on that.

The fact that the government is allowed to steal my money and incarcerate me if I don't comply, so that they can use it for whatever it feels like—like a good-for-nothing study, or an anti-vaping campaign, or a war—is infuriating. They take our money to spend on wars that they lie to us about, and then gaslight us whenever we happen to notice that they've been lying, and then they do it again. An example: *The Afghanistan Papers* revealed that our military leaders routinely told us that things were going great over there even though they knew that they weren't—and we have yet to see any accountability for the lies or losses of life that we witnessed during that decades-long disaster. Reading this book, you'll learn a lot more about my dating life than we'll ever learn about any corrupt government operation, even though my dating life has wasted far less money than any of them—and that's despite the fact that I've dated so many broke, unemployed losers that when Lou Reed sang, "Oh baby can I have some spare change? Can I break your heart?" I *felt* that.

Meanwhile, the government insists on locking people up for having any involvement with what they've deemed to be Bad Plants (even those with documented medical uses, like cannabis or psilocybin mushrooms) or for a consensual business transaction between adults, if that transaction happens to involve genitals. (They *own* our genitals, man!)

If it wasn't clear, I'm a libertarian. I believe in both the Second Amendment and that, in terms of immigration, any nonviolent person who wants to come to the United States to contribute to our economy should be free to do so. This doesn't give me any choice but to vote strictly libertarian down the ballot any time that I've voted—and, when there's no libertarian option, to write in my cat's name. Yeah, Cheens Timpf has been nominated for a lot of judgeships in the state of New York. Yes, I've certainly gotten a lot of grief for this, and have faced my fair share of accusations that I'm wasting my vote, but I just don't see any other choice. Both major parties stand for things that I don't agree with; both major parties steal from me through taxation; both major parties have used that money to fund things that I don't support, and both major parties support laws and restrictions that I believe violate my individual rights as well as the rights of others. To me, voting for either one of them would feel like saying that I'm okay with all of this—that I consent to my rights and my money being stolen from me. If you see it as a binary choice, that's fine. You can do whatever you want with your vote. Just please show me the same respect when it comes to mine.

Anyway, growing up, I saw this as my dream career not only because I loved comedy and hated politics—I wasn't even old enough to really know what love-hate could translate to in terms of sex—but also because of the way that comedy can add a disarming quality to difficult discussions. Jokes and humor can make it easier to navigate those conversations and calm down the hate.

Adding some humor can be the best way to get your points across, or, at the very least, to get people to listen. Think about it: Not everyone is going to agree with whatever you might say, but everyone enjoys laughing. There is, after all, a reason why Comedy Central tends to have more viewers than C-SPAN, and it is not because there are people who have not yet seen *Wedding Crashers*.

What's more, a 2021 study published in the *Journal of Communication* found that using humor in political discussions made it far more likely that people would remember and share the information. An article in *Forbes* about the study explains:

> *Results showed that humor increases your attention because you have to follow the thread of the joke. Anything that demands your attention also increases your likelihood of remembering. The reward response to humor comprehension, "getting the joke" also helps make information more memorable.*
>
> *The desire to share humor with others is based on increased activity in regions in the brain involved in trying to understand the mental states of others.*
>
> *The researchers lay the groundwork for a theory that humor may help us take into consideration other people's views. Humor may also de-stigmatize political conversation, making politics more palatable to people who don't like thinking about or discussing politics.*

To be fair, the study did find one issue with political satire: According to the lead author, Jason Coronel, an associate professor of communication at Ohio State University, "people actually have a hard time figuring out whether satirical information is true or false."

To an extent, I get that; our politicians have behaved in some truly hard-to-believe ways. For example, I will never really be able to wrap my head around the fact that Senator Elizabeth Warren, a white lady from Oklahoma, *really did* spend decades of her life pretending to be Native American—indicating it on her State Bar of Texas form in 1986 and contributing recipes to a family cookbook literally fucking called *Pow Wow Chow.* Republican representative Marjorie Taylor Greene *really did* post a ton of deeply insane things on her Facebook account a few years before being elected—including an agreement that the mass shooting in 2018 at Marjory Stoneman Douglas High School in Parkland, Florida, had been staged and a suggestion that the 2018 California wildfires had actually been started by Rothschild-funded space lasers. (No, I'm not saying the two things are the same, simply that they are both real assertions, despite seeming too bizarre for that to be the case.)

I could give countless examples, but you get it. The point is, when you have real politicians actually doing and saying things like that, it can be hard to distinguish reality from satire. You could play "American Politics or Something a Tweaked-and-Toothless Man Mumbled to Himself on the Train" and have an extremely difficult time figuring out which was which—unless you had been paying close enough attention to know all of the facts about those nuts-but-real political stories.

Oddly enough, a lot of research suggests that the solution to the reality-or-satire problem, as well as the larger issue of political engagement, could also be comedy—because political comedy encourages people to pay more attention to politics and to think more critically about the issues than they would have otherwise.

Consider a paper published in the *Review of Communication* in 2013. The authors, Amy Bree Becker, assistant professor at Loyola University Maryland, and Don Waisanen, associate professor at Baruch College, summed it up this way:

> *In the area of comedy effects, we found a decade of work considering the effects of exposure and attention to late-night comedy content on*

> *democratic citizenship. One set of variables has dealt with citizens' knowledge and learning, showing that such content generally increases knowledge about politics and helps people pay more attention to elections. This research also has looked at the impact of comedic messages on people's attitudes and opinions, and the overall levels of cynicism and engagement that result. One overarching finding is that political comedy appears to promote more cynicism toward politicians, the government, and the media, but also tends to empower citizens to think they can contribute to and make a difference in politics.*

Got that? Comedy not only makes people pay more attention to politics during the joke, but also overall. Even better, it makes people more likely to be skeptical of government. It's always smart to be skeptical of those who have power over you, for the sake of your own power. It makes people more likely to doubt the media, meaning they're more likely to ask questions than automatically fall for a narrative. Ultimately, the fact that our leaders do so many absurd things so often is not an argument against satirizing them. It is an argument for doing so more often.

The *Washington Post* published a piece about how Ukraine's official Twitter account was posting comedy content about the Russian invasion. The article, titled "What's So Funny About a Russian Invasion?" listed tons of reasons, including how "satire and dark humor can help individuals counter feelings of powerlessness and distress."

That's not unique to politics or war, but many of the other reasons were:

> *The use of humor in politics has been around for a long time. Many scholars have noted its power to puncture vanity, expose hypocrisy and challenge falsehoods. This makes it perfect for provocations*

> *and attacks on authorities. In autocracies, however, rulers are often protected from direct disparagement. They react harshly against people who publicly make fun of them. Recent research on civil resistance elaborates on this subversive potential and shows that humor can persuade the public that repressive tactics are ridiculous and excessive.*

It's absolutely true, and Ukrainians are far from the only people who have made use of it. Britain, for example, was able to reach German civilians with satirical radio shows during World War II. A piece in the BBC notes the unique advantages to speaking truth to power through comedy, "especially for those under tyrannical rule; comedians can claim they were just kidding, after all, or subtly mock a leader without naming him or her."

Even in a country where we do have the First Amendment, though, comedy still provides a unique opportunity when it comes to keeping our political leaders in check. Some of the power that any leader has, after all, comes from the fact that we *know* that he or she is a leader with power, which implicitly suggests that he or she should be treated with a certain level of decorum beyond that of the average citizen. If you think about it, Being a Powerful Person is a sort of power in itself. Simply because of your position, people can see you as being on some kind of higher level—making it harder for them to realize that you're capable of mistakes, stupidity, or worse.

Unfortunately, viewing our leaders as being somehow better than we are is among the best of ways for them to take advantage of us. A lot of people in power abuse it, and even those with good intentions might not necessarily be informed enough to be able to make our decisions for us better than we could for ourselves. Mocking them, though, chips away at the facade that they are anything but people just like we are. It brings them down to our level, making us feel more comfortable that we're qualified to question their motives and decisions.

Comedy in politics is important. So important, in fact, that we have to make sure we keep the politics out of comedy—and the rest of our speech as well.

In 2017—does anyone remember those pre-pandemic years?—I took the side of comedian Kathy Griffin, who got effectively canceled, put on the no-fly and Interpol lists, and was subject to a two-month-long federal investigation after holding up a ketchup-covered, fake decapitated head of then-president Donald J. Trump. To be fair, I didn't find it funny at all, but in a column for *National Review* at the time, I explained that, although I totally understood why people were upset by it or disgusted, there was no reason for Griffin to be criminally prosecuted. I wrote:

> *In order for something like this to qualify as a criminal threat, it has to be a "threat to take the life of, to kidnap, or to inflict bodily harm" that was "knowingly and willfully made"—and no reasonable person could possibly think that this is the case when it comes to Kathy Griffin.*
>
> *This was not a threat. This was a desperate plea for attention coming from a woman who has made a career out of desperate pleas for attention, and who is now finally as famous as she has spent her whole life trying to be.*

Given the audience of the network that I'm on, I'm sure you can guess that not everyone totally agreed with my take—either in that column or on television. Social media at the time was full of people remarking that Griffin *should* be prosecuted.

I still believe now what I believed then: Griffin is a comedian who pushes boundaries. I mean, damn, she's certainly gone for the super-controversial before. In 2007, for example, she accepted her "Best Reality Series" Emmy for *My Life on the D-List* by saying: "Suck it, Jesus, this award is my god now." I mean, guys? It was always so clear that the star of *My Life on the D-List* was not going to murder the president of the United States; I was shocked that it was even taken as seriously as it was. It went too far; it misfired; it was gruesome; it was at the expense of the president, and it was all protected speech. And that's a wonderful thing. We can't afford to take that for granted. As the photographer, Tyler Shields, put it: "[I]f this was a different time or a different country, I'd probably be killed right now for these photos."

Both Griffin and Shields have explained that the photo shoot was intended as a play on Trump's "blood coming out of her wherever" comments about Megyn Kelly. Our constitution protects speech, including artistic expression, not *despite* the fact that some of it might be critical of our country's government, but specifically *so we can* be critical of our country's government without retaliation from it. That's the purpose of the First Amendment!

Unfortunately, a lot of people are unable to see this. A perfect example, I'd say, would be liberals since Trump. (You didn't think I was only going to give examples of *Republicans* opposing free speech, right?)

People on the Left call for limits to speech all of the time, and for things far less on-the-nose incendiary than a comedian holding up a bloody replica of the president's head.

Hillary Clinton, for example, has called for governments to regulate speech on Facebook to prevent the spread of "misinformation." Worse, Democrats actually in government positions of power (sorry, Hillary) have made moves to censor online speech, using that same subjective term, "misinformation." In July 2021, for example, then–White House press secretary Jenn Psaki straight-up admitted that Joe Biden's White House was "in regular touch with the social media platforms" to flag "misinformation, specifically on the pandemic."

The problem here, of course, is that what the government may *deem* "misinformation" doesn't always turn out to be misinformation at all. Remember, for example, the Wuhan, China, COVID lab leak theory? The thing that went from a whacked-out conspiracy theory to a plausible hypothesis? Or how Hunter Biden's laptop turned out to be Hunter Biden's laptop? I always knew it was his, because, no matter how many people denied it, Hunter was never one of them. Like, if you *really* weren't smoking crack with escorts, you'd probably want to come out and make that clear. Generally, if you're accused of something outrageously, salaciously awful, and you didn't do it, wouldn't you? Sometimes people repeatedly deny it even when they totally did do it. (I'm looking at you, Joy Reid.)

The government trying to make rules on censoring "misinformation" or "disinformation" just means that the government gets to define what that is. It's always a changing target, and easily abused for political purposes. Plus, even if the best of intentions were always maintained, it would still go wrong because the government is made of flawed humans who fuck up. Hell, government agencies *themselves* post things that turn out to be wrong all the time—so, basically, when anyone calls for the government to censor "misinformation," what they're *really* saying is that government officials and agencies should be the only ones permitted to dispel it.

Still, in 2022, the Biden administration even announced the launch of a Disinformation Governance Board, led by Nina Jankowicz—a person who had literally spread disinformation regarding the Hunter laptop and praised Christopher Steele as a disinformation expert. A disinformation board led by a disinformation spreader! I've said this before, but I'll say it again: I get that everything is polarized these days, but I still do not understand how we can all not agree that this is a batshit, awful move. If your reason for being cool with it is that you are a Democrat and this is a Democratic administration, then there really are no words for how stupid you are. Once an administration creates something, after all, it usually sticks around; it's not like the government likes giving up its own power. (Patriot Act, anyone?) People defending the board at that time should have realized that the administration won't always be Democratic, and that when it's not, the board will likely still be there—and some of the people who were cool with it at first might not be *quite* as cool with it once, let's say, the DeSantis administration uses it to declare that it's disinformation to consider anyone without a vagina a woman.

The Biden administration did decide to "pause" the board just three weeks after launching it. Jankowicz resigned, and explained the reason for the pause in an interview on MSNBC's *All In with Chris Hayes*, saying, "Every characterization of the board that you heard up until now has been incorrect, and frankly, it's kind of ironic that the board itself was taken over by disinformation when it was meant to fight it."

In the interview, Jankowicz insisted that the board was never intended "to say what was true or false," but simply to "equip people with the tools" to spot disinformation, and that any issues that people had with the board were simply due to a right-wing smear campaign against it. A lot of the media had her back in terms of this. The *New York Times* ran a piece with the headline "A Panel to Combat Disinformation Becomes a Victim of It." A *Washington Post* piece by (who else!) Taylor Lorenz called "How the Biden Administration Let Right-wing Attacks Derail Its Disinformation Efforts" might as well have been done by Jankowicz's PR team, it was so absurdly one-sided—touting Jankowicz's "extensive experience in the field of disinformation," bragging that "her work was well-regarded" among "disinformation researchers," lamenting the fact that "her role [was] mischaracterized as she became a primary target on the right-wing Internet," and blaming the board's and Jankowicz's demise on "far-right influencers." Throughout the entire 1,994-word piece, there is not even a single mention of how Jankowicz had herself spread disinformation when it came to the Hunter Biden laptop story. To me, an omission like this actually renders the whole article false. By refusing to include such an obviously legitimate reason for criticizing Jankowicz, it gives the reader the impression that anyone who had any issue with either Jankowicz or the board must have only felt that way because of the sinister work of a right-wing lies-and-hatred machine. Lorenz left out such an important part of the story, she turned the story into something else. Put another way, her article was the journalistic equivalent of a guy who runs around talking about how his crazy ex-girlfriend once ran into their bedroom and started throwing things at him, while failing to mention that his ex-girlfriend's naked roommate had been in the bed with him, too.

Again, establishing a Disinformation Governance Board inherently implies that the government is the arbiter of truth. Put another way, creating a government agency to separate truths from lies communicates that it is the government's role to tell us what to think. Its very existence separates government-sanctioned speech from all other dissenting opinions. It's definitely disturbing, and should be unacceptable. The government's words, after all, should always be taken with the understanding that it may also have the motivation and means to spin—and, of course, the knowledge of how frequently the government has already been busted lying to us in the past. (Damn, I can't believe we still haven't found those weapons of mass destruction.) All of this is especially worth noting, considering the findings of an investigation conducted by

*The Intercept* that were published in an October 2022 article in the publication, stating: "Though DHS shuttered its controversial Disinformation Governance Board, a strategic document reveals the underlying work is ongoing."

Another leftist push, of course, was the call for the government to regulate so-called hate speech.

"Hate speech isn't free speech" is a common attitude, of course, on college campuses—you can read more about them and their methods in my chapter discussing the idea of words as violence—and was even the inspiration of a bill proposed by a New York state senator at the start of 2020, which sought to ban "hate speech" from social media platforms.

Honestly, one of the most mind-bending observations about recent years has been how many people simultaneously hold the view that hate speech needs to be shut down *and* the view that Donald Trump is Literally Hitler.

In 2019, for example, former *Time* editor Richard Stengel wrote in the *Washington Post* about how the United States needs a federal law banning hate speech, and that Trump would be in violation of it. After I stopped laughing, I wrote a piece in *National Review* titled "Former *Time* Editor Wants Hate-Speech Laws, Thinks Trump 'Might' Violate Them, and Misses the Irony," explaining:

> *It's interesting how Stengel actually does acknowledge the fact that "there's no agreed-upon definition of what hate speech actually is," and yet he still wants laws banning it. This makes absolutely no sense. After all, when he calls for laws to ban "hate speech," he is, inherently, giving the government the power to decide what would and would not qualify—the exact same government that is led by Donald Trump, and that is full of people who support him.*
>
> *In other words: Stengel somehow trusts that the government will have the same view of "hate speech" as he does, and then, in the*

*same thought, seems to acknowledge that there's actually no way that many of them would. Unless he thinks that the president and his congressional supporters would actually pass a law that they'd be in violation of, his argument for "hate speech" laws winds up being a pretty great argument against them.*

*It's ironic, but it's not new: More often than not, it's the uber-progressives arguing for laws against "hate speech"—despite the fact that they're often the same people who are also arguing that Donald Trump and Republicans are constantly spewing it. Maybe it's just me, but if I thought that the leader of my government was, you know, literally Hitler or whatever, the last thing that I'd want would be to give that person and their supporters control over my speech.*

It's the problem with hate speech laws in general—there is no clear definition of what that means. "Hate speech" is in the eye of the beholder, yet a shocking number of Americans seem to believe that the government should be that beholder—or, worse, believe that it already is. According to a poll published in 2021 by the nonpartisan Freedom Forum, only 57 percent of Americans polled even knew that hate speech was protected speech under the Constitution, 24 percent said hate speech should be outlawed, and 36 percent said preventing hate speech is more important than preserving free speech.

It's far from a majority, sure—but it's still an absolutely terrifying number when you consider what we would be giving up if anything were to change.

It's not hard to see what we would lose in terms of comedy. Think, for example, of the pushback that Dave Chappelle faced after he made jokes involving gender identity in his Netflix special *The Closer.* For example, Imara Jones, the founder of a transgender-focused media nonprofit called TransLash Media, also known as the host of the first high-level United Nations meeting on gender diversity, told (taxpayer-funded) PBS:

> *Well, I think the first thing to realize is that this is essentially hate speech disguised as jokes.*
>
> *And that is an essential point here, that no one is contesting that humor can be outrageous, sometimes offensive. But I think that this crosses the line into hate speech that's disguised as jokes.*

Jones states all of this objectively, as if it's a fact. Of course, it isn't. Many people did not consider Chappelle's jokes to be hate speech—including the family members of Chappelle's late trans friend Daphne Dorman, whom he mentioned in the special. Her sister, in fact, referred to Chappelle as "an LGBTQ ally."

Those who have a different opinion on Chappelle and his jokes than Jones don't simply, as she suggests, fail to "realize" it. It's just that it isn't an objective standard. Without an ironclad First Amendment covering everything, including hate speech, Dave Chappelle could have wound up in jail.

Outlawing hate speech may sound nice and warm and fuzzy—until you realize that all you're doing is giving the government the power to decide what kind of speech is and isn't permissible. If you absolutely hate(d) Donald Trump, if you consider(ed) him an evil monster hell-bent on destroying our democracy, then you should have spent the entirety of his presidency being grateful for those rights. It is, after all, exactly because of them that when Trump tweeted that the amount of time *Saturday Night Live* spent making fun of him "[s]hould be tested in courts, can't be legal?" he didn't actually have any legal options for punishing those comedians. As with Chappelle, without a strong First Amendment, the government would have had an opportunity to find a way to classify those sketches as hate speech. When it's up to the people in the government, it can be whatever they want. That's how "up to them" works!

Now, to be fair, I can only sympathize with what it's like to want to shut down speech due to an allegiance to a major political party—as a small-*l* libertarian, I don't have the tribal urge to shut down speech from the Other Side that I'd excuse from my own.

I do know what it's like to stay consistent on keeping speech free even when you hate what's being said, because I've explicitly defended speech that was brutalizing *me*. I mean, people have made horrific comments to me on the Internet. Creepy stuff about wanting to cum on my glasses, mean stuff about how I should kill myself or I'm stupid and useless, or I'm so awful I'm probably the real reason my mom is dead, or that I'm "an embarrassment to the human race," or defending the Nazis on the grounds that our culture would be better without Polish Catholics like my family and me.

It can really suck to see cruelty like that—but, despite it sucking, I once wrote a piece in *National Review* titled "I'm the Target of Hatred, and I'll Still Defend It as Free Speech." The truth is, nothing anyone could ever say to me would ever be as upsetting or terrifying as the thought that my own right to speak as I choose could ever be taken away. If you still think words are violence, picture yourself being tackled, cuffed, and shoved in the back of a police cruiser for saying that Kat Timpf is a 6 without her glasses and hair extensions.

Many of the people who want to censor speech see their belief as a form of compassionate devotion to marginalized groups. Those intentions are good, but the truth is, marginalized groups are the exact people who have historically benefited from free speech the most.

As Jonathan Zimmerman wrote in the *Wall Street Journal*:

> *When speech can be suppressed, the people with the least power are likely to lose the most. That's why every great tribune of*

*social justice in American history—including Frederick Douglass, Susan B. Anthony and Martin Luther King Jr.—was also a zealous advocate for free speech. Without it, they couldn't critique the indignities and oppression that they suffered.*

The examples of this are countless—from abolitionists in the antebellum years to gay publications in the mid-twentieth century, it was free speech and the First Amendment that thwarted the censorship of ideas of a marginalized minority that many people in power considered to be offensive or wrong. When anyone calls for a ban on hate speech, they're assuming that the people in charge of that are going to have the exact same definition of what that means as they do. I hate to say "slippery slope," but that slope really is mad slippery. Hearing something awful can suck, but the best answer to speech you don't like is always to respond to it with . . . your own speech. It's great fun. Give it a go. I'll start: If the cool kids at Wellesley College had just let Hillary Clinton sit at their lunch table, none of this would have ever happened.

CHAPTER 12

# FREE SPEECH AS A CULTURAL VALUE

By the time you're reading this, society may have already devolved into a dangerous, hate-crime-ridden wasteland because Elon Musk bought Twitter.

After all, when the deal was first announced in April 2022, the predictions for what might happen were downright dire. Representative Alexandria Ocasio-Cortez said it would cause an "explosion in hate crimes." A group of twenty-six organizations including Media Matters for America and the Women's March signed a letter telling advertisers to boycott Twitter if Musk maintains his free speech–first vision, because that vision would "provide a megaphone to extremists who traffic in disinformation, hate, and harassment . . . silence and endanger marginalized communities, and tear at the fraying fabric of democracy."

If I had taken those warnings to heart when I started this chapter, I would have probably had to ask myself: Damn, should I even *keep* writing? By the sound of all of that, we're going to be living in a racist, sexist edgelord-archy, where women are hunted down and shot for sharing their opinions, before this book ever even has a chance to come out. I hope I'm not an idiot to just calmly sit here typing after I've been warned of a coming apocalypse—like one of those college kids in a horror film who decide to keep driving to the lake cabin even after an ominous experience at a run-down gas station on the way. Especially since the blond bitch usually gets killed off first.

Of course, Musk had prompted those fears by having touted himself as a guy who was "against censorship that goes far beyond the law." Once he actually took control of Twitter in the fall, however, it started looking like the concerns about him turning it into unbridled-speech free-for-all might not have been warranted—if only because he started suspending people for "impersonating" him by changing the names and photos on their accounts to look like his and then tweeting. Kathy Griffin's suspension in particular struck a nerve because she is a professional comedian and was pretty clearly just trolling and looking for attention more than earnestly hoping to run an account that people would believe belonged to Musk. (Especially because, just days before her suspension, Musk tweeted, "Comedy is now legal on Twitter.")

But I will get into Musk vs. Griffin later. First, I want to talk about how Elon Musk offering to buy Twitter—and all of the resulting predictions of our societal demise—happened all because of comedy.

Musk, after all, only started talking about buying the platform after Twitter suspended the Babylon Bee, a satirical website, over one of its jokes.

In March, the Bee tweeted a satire piece about Rachel Levine, the transgender assistant secretary for health for the US Department of Health and Human Services, with the headline "The Babylon Bee's Man of the Year Is Rachel Levine." Twitter then notified the

Bee that it had locked its account because the tweet had violated its policy against "hateful conduct."

"You may not promote violence against, threaten, or harass other people on the basis of race, ethnicity, national origin, sexual orientation, gender, gender identity, religious affiliation, age, disability, or serious disease," the notice read.

Shortly after that, Musk polled his followers, asking for their views about Twitter's commitment to free speech. And it wasn't just a coincidence: Later, the Bee's CEO, Seth Dillon, would tweet that Musk had indeed called him to confirm that his publication's account had been suspended for that joke before posting the poll. Twitter had confidently imposed that suspension in a show of power over the Bee, but it actually wound up being the catalyst for the entire power dynamic shifting.

The craziest part? The joke wasn't even good. It certainly wasn't creative! I say this as someone who finds a lot of the Bee's stuff to be absolutely hilarious ("AOC Says She Got Killed from Elon Buying Twitter and Is Now Dead," "Trump Rescinds Pardon After Learning Turkey Immigrated from Mexico," and "CNN Apologizes to Stalin, Mao After Comparing Them to Trump," to name a few) but the Rachel Levine tweet was pretty low-level hack. *Oh, this person who identifies as a woman was born with a penis. Ha ha! Get it? She's a dude!* Like . . . eh. Even Reddit would have downvoted that, and some of the worst people in the world are on Reddit. I would know; I'm lurking there every day.

Of course, Twitter had every right to do what it did. But you don't have to like the joke to realize that it clearly did not amount to *threatening* or *harassing* or *promoting violence* against someone the way that the Twitter rules alleged that it did. A single trite joke does not amount to any of those things—and honestly, the fact that it was so trite is pretty good proof of that—making a suspension on those grounds completely ridiculous. Hell, I've seen trolls in my own mentions allege that I was born a man—complete with a diagrammed analysis of my facial structure as proof of the claims—and I never thought that those people should be suspended, either. I have bigger things to worry about, plus I know what it says on the birth certificate that I keep in a shoe box in my closet.

But the main reason I felt bothered by the suspension was this: I think that speech is good, and I want to stand on the side of it.

Just as in the case of the old Twitter leadership and the *Bee*, Musk-owned Twitter had every right to suspend Griffin. What's more, Twitter's policies at the time of Griffin's suspension clearly stated that users "may not impersonate individuals, groups, or organizations to mislead, confuse, or deceive others, nor use a fake identity in a manner that disrupts the experience of others on Twitter," meaning that Griffin's suspension may have made more sense than the *Bee*'s did if you look at it only in terms of the rules. Still, rules aside, Musk suspending Griffin did not seem to fall in line with the free-speech absolutism that he himself had said would guide him in governing Twitter. This would have been especially true if Musk's ban of Griffin had turned out to be permanent, as an initial report had stated that it would be. Suspending someone temporarily for impersonation-trolling is one thing, banning them forever for impersonation-trolling is quite another. The latter would have been quite clearly ridiculous and unfair, and especially absurd and hypocritical coming from someone who has claimed he values free expression and comedy. I am glad that, at least at the time of this writing, both Griffin's and the Bee's accounts have been reinstated.

The bad news is that I have no way of knowing how Musk will be running Twitter by the time this book comes out. I don't visit psychics! I can admit that I've wanted to try it, but I'm not convinced that they know anything, and it doesn't seem worth the risk that they might blab about my life to whoever they felt like because there's no psychic-client privilege and I can't believe that no one seems to worry about that like I do. The good news is that I have principles that transcend tribalism, and that, whatever he does or does not do, my view of his actions will be based on those principles alone. (This is despite the fact that, at least at the time of this writing, my own Twitter follower count has skyrocketed since Musk announced he'd be buying it. Many of my Fox News colleagues have had a similar experience, making me believe that the old Twitter leadership seemed to believe that my employment told them everything that there was to know about me . . . which is quite an ignorant assumption.)

In my political chapter, I focused on the First Amendment: our right to protection from government retaliation for speech, which is both gravely important and completely different from what I'm talking about here.

Or, as my friend and former congressman Justin Amash tweeted in the wake of Musk bidding for Twitter:

> The First Amendment and free speech are not completely overlapping circles in a Venn diagram, but people often mistakenly treat them as synonymous. The error runs both ways. Some think any free speech issue is a 1A issue, while others think only a 1A issue is a free speech issue.
> Put another way: There's a difference between the government *violating* someone's *right* to free speech and people or a company *not upholding* the *ideal* of free speech. The former is a First Amendment issue; the latter is not. Both are matters of concern in a free society.

Whether or not any given shutdown of speech counts as a First Amendment violation can be a valuable discussion to have, but determining that one did not count is also not a sufficient justification for not caring that it happened.

For one thing, making someone shut up is not the same thing as making him agree with you. If anything, silencing people can make them even more passionate about their "forbidden" opinions. When you do that, it just solidifies your position as the unreasonable, domineering enemy, and theirs as the victim and martyr. It might discourage them from ever wanting to engage with anyone with your opinion ever again. It's the equivalent of telling someone who's upset to "calm down": It only makes it worse. Strict cultural expectations about speech don't help us to get along. They prevent us from having the exact conversations we would need to have to get there.

In March 2022, Senator Marsha Blackburn of Tennessee asked Ketanji Brown Jackson to define the word "woman" during Jackson's confirmation hearing for the US Supreme Court, and Jackson infamously replied: "I'm not a biologist."

It was a cartoonishly stupid answer. It might have been one of the worst answers to a question that I've ever heard, maybe even including some of the ones that I've heard on *Cops*. (Why do all of the blitzed-out, covered-in-crack-dust people always claim that they've had exactly "two beers" when the cop pulls them over? It's never "one," it's never "none," it's never "wine," it's always exactly "two beers." And the pipe in the glove compartment is always an immaculate conception!)

Pretty much everyone mocked Jackson. Social conservatives mocked her because their definition of a woman was a matter of *simple* biology: a person with two X chromosomes. On the other side, a piece in *USA Today* explained that gender experts felt the problem with her answer was that the definition of a woman was *not* a matter of biology at all:

> *"I don't want to see this question punted to biology as if science can offer a simple, definitive answer," said Rebecca Jordan-Young, a scientist and gender studies scholar at Barnard College whose work explores the relationships between science and the social hierarchies of gender and sexuality.*
>
> *"As is so often the case, science cannot settle what are really social questions," she said. "In any particular case of sex categorization, whether in law or in science, it is necessary to build a definition of sex particular to context."*

I was one of the few people who did *not* mock Jackson, and that's even though mocking people is generally a pretty big part of my job. When we talked about it on *Gutfeld!*, I explained that I could actually understand what might have caused her to answer that way.

See, just days earlier I'd had to talk about transgender University of Pennsylvania swimmer Lia Thomas on the show. I was terrified from the moment that I found out I'd have to do so. Talking is my job, sure, but the idea of talking about *that* filled me with fear that I would accidentally say something that I wasn't supposed to say.

So, on that show (after agonizing about it for almost a full twenty-four hours, including one text fight with Gutfeld about how much I did not want to talk about it), I decided to go with just being honest about everything—including *and especially* the turmoil I'd been experiencing about the whole thing.

When Greg went to me, I admitted that I had been really nervous to talk publicly about the subject, and that I kept catching myself thinking that I'd much rather talk about the literal war in Ukraine than this college women's swimming competition. To me a murderous, devastating war felt like less of a touchy subject, and those feelings ultimately made me realize that there must be something fundamentally wrong with our collective discourse on this issue.

I acknowledged, of course, that transphobia is awful, that there's no doubt that Lia Thomas has had to face it, and that that was wrong. Still, it's also a fact that USA Swimming's new standards for trans athletes would have likely disqualified Thomas from competing based on her testosterone levels if she had been required to meet them, and simply pointing something like that out doesn't equal transphobia:

> *So, there are legitimate reasons to have questions about this that have nothing to do with transphobia. And I think it's bad to just blanket-label anyone who says anything short of "You go, girl!" as transphobic, not just for the biologically female athletes, but also for people who want trans acceptance.*
>
> *Because there are a lot of people who have questions about this, who have concerns about this, think maybe this wasn't fair, who are not allowed to say that. And that is not acceptance. If anything that can prevent acceptance, because it breeds a lot of resentment.*

That's what I said, and do you know what? I agree with me. Even in cases where it may be well-intentioned, making people terrified to talk about something just won't be the most effective way to promote the kind of understanding and acceptance that you claim to want.

I don't know Justice Jackson, but honestly, I would guess, based on how absolutely awful her answer was, that this is exactly what had happened to her. You do not, after all, get to a Supreme Court confirmation hearing because you're a dumbass. To me, it seems far more likely that Jackson had heard a question related to gender, and alarm bells started ringing in her head: *Danger! Danger! Approaching trans stuff! Risk of cancellation ahead!* That has certainly happened to me—including during that Lia Thomas segment, actually. The reason I explained some of my answer in this book rather than just copying and pasting the entire thing verbatim was that part of it was a little mumbly due to my nerves. I wanted to make sure I got it right, even as I was explaining how harmful the pressure to "get it right" is for every single one of us.

The stakes for making sure you "get it right" these days are, unfortunately, extremely and oppressively high. What's more, it's no surprise that comics ranging from Dave Chappelle to Kathy Griffin have been expressing concern about it for years, nor that something tied to comedy was what made Musk ultimately offer to buy Twitter for the sake of it.

As Jordan Peterson explained during an interview with podcaster Dave Rubin:

> *What's worrisome about the state of discourse in the free West is that comedians won't go to university campuses. It's the same thing. You don't get to be funny. So, if you can't be funny then you're not free. You know, the jester in the king's courts is the only person who gets to tell the truth and if the king is such a tyrant that he kills his jester, then you know that the evil king is in charge. So, when we can't tolerate our comedians, it's like, "Well, there you go!" They're the canaries in the coal mine as far as I'm concerned.*

Even if Peterson *hadn't* said lots of other things during this interview that made me believe he would totally love this book—such as "We need to be able to be horribly funny, because life is horrible, and we need to be able to find, we need to be able to allow people the freedom to find the ability to transcend that horror with comedy," I would still have to say that I totally agree with him here. (I say "here" because I don't agree with him on everything. I do not and probably never will, for example, make my bed every morning. I'm lucky if I manage to not leave a wet towel on top of it. Yes, I am a monster, and have learned throughout the years that it is impossible for people to cope with living with me if they are not already in love with me. Actually, I'm shocked that it has not yet dissuaded my extremely neat and tidy husband from *continuing* to love me. Although, the first time he came over, there was a pasta sauce stain on the floor, and he asked me what it was, and I said, "I don't know, I think ziti?" And he was like, "I wasn't asking *what kind of* pasta," and actually paid for a cleaning person to start coming to my apartment every other week so he could feel comfortable even being there. So what I'm saying is, like, he knows who he married.)

I'm not the only person who agrees with Peterson on this stuff.

When Jon Stewart accepted the 2022 Mark Twain Prize for American Humor, he said something similar:

> *Comedy doesn't change the world, but it's a bellwether. We're the banana peel in the coal mine. When a society is under threat, comedians are the ones who get sent away first. It's just a reminder to people that democracy is under threat. Authoritarians are the threat to comedy, to art, to music, to thought, to poetry, to progress, to all those things.*

Of course, I say "similar" and not "the same"—because, while Peterson routinely acknowledges the way that cultural pressures can chill free speech, Stewart remained focused on the government as the True Threat, saying, "It's not the fragility of audiences. It's the fragility of leaders."

Stewart talked about Bassem Youssef, an Egyptian surgeon who hosted a satirical show inspired by Stewart's *Daily Show* during the

Arab Spring—using humor to speak out against the lies and injustices of the tyrannical government, until he was forced to flee the country out of fear of arrest for having done so.

Is a story like Youssef's a cautionary tale? Yes, of course it is—and that's exactly why I'm so passionately against giving our government even the slightest allowance when it comes to controlling speech.

But why can't we be concerned about both? And I'm not just saying that because I have anxiety disorder and have been chronically concerned about almost everything since childhood. Ask my parents, whom I would routinely wake up in the middle of the night in the third grade to ask them if I was going to have any long-term health problems because I'd only had four glasses of water that day when I had read somewhere that I was supposed to have eight.

Or, put another (more positive!) way, it's because I see the value in both protections from government retaliation against speech, *and* a society where citizens celebrate free, open conversations as a cultural ideal.

Elsewhere in his interview, Peterson says that comedians are supposed to be "people who push the edge of what's acceptable." That's true, and inherent in that truth lies a major reason why all of us should be keeping a close eye on what risks they are and are not willing to take: The less they're willing to push it, the closer that "edge of what's acceptable" gets for all of us, comedians or not.

Even if you can't be jailed in this country (yet) for crossing a line, knowing that you can face other consequences—like ostracization, cancellation, and harassment—can still have a chilling impact on speech. Unfortunately, the immense benefits of comedy, especially the kind of comedy that flies in the face of the sacred, are threatened when people become too afraid to take risks, whether what they're afraid of is jail or not.

We have, after all, already seen changes in what we will allow from comedians, and especially in what consequences they can expect for running afoul of those expectations.

In 2018, Peterson shared an article about comedian Norm Macdonald's should-have-been press tour for his upcoming Netflix show, *Norm Macdonald Has a Show*, turning into an "Endless Apology Tour" over a series of mistakes.

It happened like this: In the wake of Louis C.K.'s masturbation misconduct scandal and Roseanne Barr's racist tweet scandal, Macdonald talked with the *Hollywood Reporter* about how he felt for them having to deal with "losing everything in a day."

"Of course, people will go, 'What about the victims?' But you know what? The victims didn't have to go through that," he said.

The controversy over his comment got so big that Macdonald's *Tonight Show* appearance that night was canceled. The same day, he tweeted an apology:

> Roseanne and Louis have both been very good friends of mine for many years. They both made terrible mistakes, and I would never defend their actions. If my words sounded like I was minimizing the pain that their victims feel to this day, I am deeply sorry.

It didn't end there. That very next day, when Macdonald was on *The Howard Stern Show*, he managed to make things worse by saying, "You'd have to have Down syndrome to not feel sorry" for sexual harassment victims.

"Down syndrome," Macdonald added, displaying the typical yes-anding tendency of a comedian. "That's my new word."

As Macdonald put it in an interview on *The View* later that week: "It's always bad when you have to apologize for the apology."

Elsewhere in the segment, he took a more serious tone, explaining what had happened on *Stern* and how awful he felt about all of it.

"There used to be a word we all used to say to mean 'stupid' that we don't use anymore," he said, "and stupidly I was about to say that word and then stopped and thought [about] what's the right word to say," noting that, as soon as the words "Down syndrome" escaped his mouth, he knew that what he'd said was "something unforgivable."

Now, to be clear, the stuff that Norm said definitely wasn't cool, and it's understandable to not love it. That's obvious, but looking closely, it becomes just as obvious that he hadn't intended to be as hurtful as he was. He was grappling with the swift cancellation of two longtime friends, realized he might have sounded insensitive, and apologized. Then, in attempting to add some off-the-cuff levity to another apology, he said something extremely offensive while trying to avoid saying something *else* offensive—and, although offense wasn't his intention, he also apologized for that comment repeatedly. He messed up, but he knew it, and his series of apologies made it clear how awful he felt about the whole thing.

Of course, none of that mattered, and Norm was absolutely dragged to Hell for the whole thing anyway. A piece in the Pacific Standard said that anyone chalking up the backlash to "oversensitivity" just did not understand that what Macdonald had really been doing was "implying that people with Down syndrome were subhuman and incapable of higher emotions." People on Twitter suggested that his upcoming show should be canceled, and an article in Salon suggested that the only reason he even had a show in the first place was institutional sexism—because Macdonald was clearly the poster boy for abuse-enabling men in positions of power everywhere, much like "that guy who holds the door open for you and buys you lunch sometimes and apologizes for his good pal Bill when Bill drunkenly pinches your boob at after-work gatherings."

We would later find out that, as this whole controversy was

happening, Macdonald had already secretly been living and working with cancer for six years. It would kill him just a few years later.

Regardless of the offensiveness of Macdonald's comments, I do think that Peterson was correct to present this as an example of the "canary in a coal mine" phenomenon—especially because of just how much it shows a massive difference between how things have changed in terms of an allowance for offensiveness from comedians throughout the years.

Think about it: Macdonald's grovelingly apologetic 2018 appearance on *The View* sure was a far cry from his *View* appearance in November 2000—where he had explained he was happy to see that George W. Bush won the White House, and just hoped "that the Democrats don't steal the election from the winner."

"I love George Bush, man, he's a good man, decent, you know? He's not a liar or crook murderer or anything," he said, adding: "I think we should get the homicide out of the White House and, like, a fresh start, because we don't want any more murderers."

When the panel expressed confusion and Joy Behar followed up by asking, "Who're the murderers?" he replied, plainly, "Oh, Clinton. He murdered a guy."

And Joy Behar? *She laughed.* I'm serious. It's an uncomfortable laugh, sure, but it's a laugh, and you can see it in a clip from the episode. She *laughed*, and *continued to laugh* as Macdonald doubled down.

Sure, the panelists, especially Barbara Walters, did push back on Macdonald's claim, but all of them did so only in a lighthearted, fun way that we rarely if ever see on *The View* these days. Walters at one point told Macdonald, "I don't want to hear it, and this is not the place to make those accusations, and you're supposed to be funny."

But Macdonald didn't back down, saying, "I thought it was a matter of record!"

Overall, it may have been one of the most hilarious moments I've ever seen on television. It was funny because it was so absurdist and unexpected, and because of Norm's matter-of-fact delivery of something wildly controversial and conspiratorial. (To be fair, this was also before the age of social media, when outrage expressed on a platform that less than a quarter of the population uses is too often used as a barometer for public opinion in making decisions about cancellations.)

When clip resurfaced in the wake of Norm's death, I couldn't help but ask myself how this same routine might be taken if he had done it these days. The obvious answer is that it would, you know, *not* be. These days, instead of laughing, I'm sure that Behar would have pointed a finger at him and accused him of spreading *dangerous misinformation*, Sunny Hostin would have demanded that Congress launch an investigation into his jokes, and much of Twitter would probably find a way to connect it to January 6th.

The day we found out that Norm died, I tweeted:

> So many comedians are scared, especially these days. Norm seemed to be the opposite: The more he felt that people didn't want to hear something, the more he felt compelled to say it—and did say it, and remembered to make it funny. There isn't enough of any of that, and we need it.

I hate to keep quoting myself like an asshole, but I was definitely right. In that same *View* segment where Macdonald jokes about Bill Clinton being a murderer, after all, he also purposely ruins a bit that producers wanted him to do with his cell phone, probably *exactly because* the producers wanted him to do it. On an episode of *The Daily Show* right after Steve Irwin died from a stingray bite while filming a documentary in the Great Barrier Reef, Macdonald insists on making jokes about Irwin's death. Jon Stewart clearly doesn't want him to, but as the bit goes on, Stewart ends up begging him to stop—because he feels bad about how hard he's laughing.

Norm seemed so devoted to joking about the exact things that he wasn't "supposed to" joke about. In doing so, he often brought laughter to the darkest topics, making us forget our sadness or fear for a moment. More importantly, seeing him speak so fearlessly could make people watching feel less afraid to talk openly and candidly themselves—even if the topic was tough, and even if we weren't sure if we had the right words—allowing us to understand each other on a deeper, more honest level. He took a lot of power away from those who aim to silence others by refusing to shut up himself.

To me, watching those two *View* appearances next to each other is a striking illustration of just how much things have changed. Again, I didn't use the #MeToo or Down syndrome examples because I thought they were inoffensive (although I do, again, believe that his heart was in the right place) but because I will never forget the shock I experienced when seeing the usually fearless Norm Macdonald's name in the same headline as "Endless Apology Tour." Like . . . it's *Norm Macdonald*!

I hate to say it, but I myself am way more of a pussy than I ever thought I would be. When Queen Elizabeth II died, I was so tempted to make a joke about it—either by tweeting "96-Year-Old-Woman Dies" or by quote-tweeting a death announcement with the words: "Omg what happened?"—but I didn't, because I was too afraid of Getting in Trouble.

Honestly, I also hate to say *this*, but part of me notices that changing times may have changed Jon Stewart as well. I hate to say it because I have long looked up to Stewart, because of the way he spent years unafraid to hit both sides. To his credit, even fairly recently he was not afraid to mock the Left and the mainstream media for having so quickly denounced the Wuhan lab leak theory, saying on an episode of *The Late Show with Stephen Colbert*: "There's been an outbreak of chocolatey goodness near Hershey, Pennsylvania. What do you think happened? Oh, I don't know, maybe a steam shovel mated with a cocoa bean? Or it's the . . . chocolate factory!"

So I'm not going to say that Stewart is all bad, or even that I don't still look up to him, because I do. But at the same time, I do notice a difference when looking specifically at his show on Apple TV, *The Problem with Jon Stewart.* Unlike the no-holds-barred Stewart of the past—the guy who was unafraid to mockingly question the Iraq War, even as most of the media showed passionate support of it, almost as if doing so was a demand of patriotism—the Apple TV show mainly sticks to woke-approved content. I don't know Stewart personally, which I consider unfortunate, but I actually really hope he is doing this because it's what he really believes, and not because he feels like he has to do it. It's pretty easy for me to respect people with whom I disagree; it's much harder for me to respect people who say things that they don't agree with themselves.

The pressure to perfectly adhere to stringent standards of speech is bad for art, comedy, and so many of the other ways that we connect with one another. In fact, a study published in the *Journal of Applied Psychology* in 2022, titled "Walking on Eggshells: An Investigation of Workplace Political Correctness," linked political correctness—even when it *was* coming from a place of sensitivity and care—to mental exhaustion that has "concerning implications, as depletion is likely to impact how well employees interact with their spouses at home in the evening."

It's also not hard to see how fear of expression could interfere with art, because art is all about expression! (At least that's what every painter/bartender in Bushwick always babbled on about when he was trying to sleep with me.)

During my time at *National Review*, I wrote about the push at multiple schools to cancel or adapt performances of *The Vagina*

*Monologues* because the show is simply not inclusive to women without vaginas. In 2019, students at Washington University changed the name of their performance to *The [Blank] Monologues* because "having a vagina and being a woman are not mutually exclusive." The school was far from the only or the first to do this: In 2015, Whitman College changed its *VM* performance to the *Breaking Ground Monologues*, and the all-women's Mount Holyoke College canceled theirs. In 2016, American University canceled theirs, too—all for the same reason.

Another school, Southwestern University in Texas, canceled its production of the play for another reason: because the woman who wrote it, Eve Ensler, is white, and the show, therefore, could never be inclusive to anyone who isn't white. For the record, although I certainly would never say I have any grasp on what it means to be a black person (in case you weren't aware, I'm white), I don't think that means that there's never been a single thing in my life that a black person could relate to, or that anything I might say about anything would automatically *exclude* black people just by the nature of me being a white person saying it.

The whole *Vagina Monologues* debacle, to me, serves as an excellent illustration of what I mean by the difference between being sensitive and respectful and taking it too far. Because I do believe that respecting trans people is important. I am extremely live-and-let-live (or, as Mase explains in "Feel So Good," "I do what work for me; you do what work for you") not only politically, but also as a way of life. You were born a man, but you now identify as a woman? Okay, cool, whatever makes you happy, and I'll call you whatever you want. What's more, I can acknowledge that being a trans person presents a unique set of challenges and experiences that I won't ever truly be able to understand because I haven't been through them, and that it's important to listen to trans people talk about those experiences in order to gain a greater understanding.

Here's the thing, though: Being born with a vagina also comes with a unique set of experiences that people born without vaginas will never truly understand. When it comes to gender identity, I understand how, as those Washington University students put it, "having a vagina and being a woman are not mutually exclusive," but guess what, kids? Having a vagina and not having a vagina *are* mutually exclusive, so what is wrong with the existence of an outlet to talk about those vagina-specific things?

An example: If you were born a man and want to identify as a woman, that's fine. However, that identification also doesn't mean that you're going to be able to relate to a joke I made on *The Greg Gutfeld Show* a few years ago when responding to a study claiming that men face worse discrimination in the workplace than women:

"Have you ever," I said, wagging my finger at Greg, "had to sit at your desk and wait for no one to be looking at you, so you can put a tampon up your sleeve and walk to the bathroom, because you can't talk about having your period . . . no, you have to *whisper* about having your period, like, 'Do you have a tampon?' like you just murdered somebody, because you have a functioning female body, and apparently, that's somehow disgusting."

But, of course, Greg *wouldn't* understand that, and neither would a trans woman. Does that mean that I should not have said it, or that I should have used some other kind of terminology except for "female body," so as not to insult and marginalize trans men who have periods?

I feel like no. I also feel like not everyone would agree with me there. I think back, for example, to a Rewire News Group opinion piece I read a few years ago slamming Iliza Shlesinger for jokes that "categorize women and men in polarizing groups while entirely disregarding the existence of anyone trans, nonbinary, or queer who doesn't fit neatly into a traditional box," as if talking about one thing somehow automatically disregards the very *existence* of everything else. I think of the push to change "pregnant woman" to "pregnant person" or "birthing person," and can't help but think about how I'm sure there's someone out there who would say that I should have instead said "menstruating body," or that I was actually completely cruel for having connected the idea of being a woman to being a person who has used a tampon in the first place.

It wasn't cruel, because the truth is, less than 1 percent of Americans currently identify as transgender—which means that the vast, vast, vast majority of people who have ever used a tampon also do identify as women, even if you combine both trans men and cis male frat boys who have soaked them in vodka and shoved them up their asses. No one deserves to be disrespected on the basis of gender identity, but I don't think that acknowledging common associations equals disrespect. Especially in terms of comedy, where acknowledging common associations in a humorous way is a pretty big part of the gig.

Just think of how many jokes would be totally destroyed, or maybe never exist, if comedians could *not* do this?

Certainly not one of my favorite jokes by Wanda Sykes, which goes:

> *There's just so much pressure on us. Guys, you don't understand! It's just—and even as little girls, we're taught, you know . . . we have something that everybody wants. You gotta protect it! You gotta be careful! You gotta cherish it! And that's a lot of fuckin' pressure! And I would like a break! You know what would make my life so much easier? Ladies, wouldn't you love this? Wouldn't it be wonderful if our pussies were detachable? Let that marinate a little bit, just think about that. Wouldn't it be great if you could just leave your pussy at home sometimes? Just think of the freedom that you would have. You get home from work, it's getting a little dark outside. You like . . . "I would like to go for a jog, but . . .*

> **tsk* it's getting a little too dark. Oh! I'll just leave it at home." And you . . . you out jogging! Yeah! It could be pitch-black, you still out there just jogging! Enjoying yourself! You know? If some crazy guy jumps out of the bushes like "AHHH!" You like *tsk* "I left it at home! Sorry! I have nothing of value on me. I'm pussy-less!"*

And she starts doing lunges, and the joke continues.

It's one of my favorite jokes not only because it's funny, but also because it's so relatable to me. The truth is, though, if you're not allowed to make connections between being a woman and having a vagina, then you're not allowed to make this joke. With this joke, it wouldn't even really be possible to just change some words around (like "little girls" to "children with vaginas") because, even aside from how creepy "children with vaginas" sounds, this joke is specifically about being a female since childhood, and the things that you hear that you have to worry about as a female child, and then the ways that you have to continue to worry about similar things after you become an adult woman—that is, presenting as someone whom a man might see and want to sexually assault because he's the kind of guy who sexually assaults women.

What Sykes talks about in that joke is, unfortunately, a common experience that anyone who has been a girl or woman with a vagina her entire life can relate to. This does not, of course, mean that trans people do not have things to worry about, or that those things do not include assault. It's not that those things don't matter. It's just that those things are not the subject of this specific joke. And guess what? That's okay.

What's more, the joke specifically makes women (I guess I should have said "people born with vaginas who still identify as women into adulthood"; it's not like that would be clunky at all, right?) actually *able to laugh* about something that, for all of us, has been a source of frustration and fear throughout our entire lives. It allows us to look back at all of that horrible, unfair terror and laugh and say, "*Wait, that* is *some bullshit, right?*" Also, because it *is* funny, it's able to make an important point in a way that everyone will be more inclined to consider, simply because it's presented in such an entertaining way.

When I talk about the way that jokes about dark, taboo subjects can have healing powers and bring people together, this is exactly the kind of joke that I'm talking about—and exactly the kind of joke that we're going to lose if we become obsessed with politically correct language at the expense of being able to actually talk to and relate to each other. Every joke won't be something that every person can relate to, but that doesn't invalidate its relatedness for those people who can relate—much less to the point that it shouldn't exist at all.

I mean, damn. I was born a woman and still identify as a woman, but that still doesn't mean that I relate to all of the jokes out there that are for or about "women," because I am also an individual person. Actually, my husband and I talk about how every TikTok (which we watch via Instagram Reels because we are mid-30s millennials and not Gen Z) from a woman clowning on all of the annoying, sloppy, lazy stuff that her husband does describes our living situation perfectly—because I'm always exactly like that husband. Gender roles do not and never will describe everyone, but that doesn't mean they can't be good fodder for living-in-Lululemon, oat-milk-drinking, stay-at-home moms to make TikToks, and do you know what? Good for them, and for everyone who relates to them, and for every single dollar worth of Bellesa and Fashion Nova sponsorships that that whole ecosystem adds to the economy.

Sometimes, when someone talks about something, it will run counter to or maybe even (gasp!) invalidate your own experience—but that doesn't automatically make it an unforgivable sin. You're not God, and you don't have that kind of power. You do have the power to speak and to share. It's true: The best part of ensuring that your own voice is heard and understood is *not* aiming to change the way other people talk about theirs; it's to talk about yours, and to encourage other people to be able to talk about it with you and learn. The answer isn't less speech; it's more.

CHAPTER 13

# TWITTER AND THE OUTRAGE MACHINE

In comedy, intent matters, the target of the joke matters, and the feeling that we're all in this together matters. Is there some place where people are all against each other, intent is easily overlooked, and the target of a joke can rally an army of humorless scolds? Of course there is.

It is a fact that Twitter is a cesspool. No matter what you have to say, you can be sure that someone on Twitter will have something else to say that ruins it. A few years ago, I tweeted a dumb joke that went, I'd like to think that I'm a good person, but know that I drink far too much La Croix for that to be true . . . only to see a reply that read something along the lines of: You should try Liberal Tears instead!

The fuck? How did you manage to make my sparkling water tweet about your own political douchebaggery? Unfortunately, it's alarmingly easy to encounter political douchebaggery on Twitter. Almost as easy as it is to get called a slut!

Twitter is also a uniquely dangerous place to share your thoughts. I've said it before, but I'll say it again: The best thing about Twitter is that not only do you not get paid for your posts, but you can also get fired for them.

Make no mistake: My sarcasm-drenched point is absolutely correct. It's true, and I'm right, and I'm also a hypocrite—because one of the places that I have made this assertion was on Twitter.

I know it's dangerous, but I can't seem to resist what a well-performing tweet can do to spike my dopamine. It's kind of pathetic, but who doesn't like that 1,000-retweet dopamine hit? It feels great. Some of these people are out here doing it for the 5-like dopamine hit. The problem, of course, is that dopamine feels *so* great it's been responsible for countless life-changing mistakes. People get so wrapped up in chasing it, they can forget what they risk leaving behind. Then, suddenly, the nanny you've been having fun banging is pregnant and—after a decade of somehow successfully hiding it—everyone finds out, and your twenty-five-year marriage to Maria Shriver is over.

Posting little jokes on Twitter is fun until one of them becomes big enough to blow up your entire life.

Bad things can happen even if you're not the one who wrote the joke. Think of *Washington Post* reporter David Weigel, who was suspended in June 2022 without pay for a month for retweeting an extremely dumb joke: Every girl is bi. You just have to figure out if it's polar or sexual. That joke was far too dumb to be worth however much money Weigel lost thanks to his suspension. The *Post* levied the suspension regardless of the fact that Weigel had un-retweeted it and apologized, explaining that he "did not mean to cause any harm."

But here's the thing: He hadn't caused any harm.

No person—bipolar, bisexual, or otherwise—was really, truly harmed by such a stupid joke. Everyone knows that, and it's also easy to guess what the real reason for the suspension was: Someone else said she was offended by the retweet. Weigel's own colleague, a fellow reporter named Felicia Sonmez, shared a screenshot of the retweet along with the comment: Fantastic to work at a news outlet where retweets like this are allowed! (Yeah, "allowed." She was literally saying, "You can't joke about that!")

People (predictably) lashed out at Sonmez after Wiegel's suspension, and some of their comments were truly awful. As I've already shared, I've been the subject of online harassment before, so I mean that sincerely—I know how awful it is. But Sonmez also seemed to enjoy making herself the subject of the story just a bit too much. So much so that it seemed like she hadn't really shared Weigel's retweet to protect anyone from anything, but to gain social currency for herself. Social currency does, after all, come with victimhood these days; no one would deny that. Plus, in the following days, Sonmez tweeted about Weigel constantly, something that no one would do if they just wanted the story to go away. That would be like begging for a bath and then going right back outside and rolling around in dog shit.

Another colleague, a reporter named Jose A. Del Real, seemed to notice the same thing, tweeting: Felicia, we all mess up from time to time. Engaging in repeated and targeted public harassment of a colleague is neither a good look nor is it particularly effective. It turns the language of inclusivity into clout chasing and bullying.

He was right. Sonmez was being extremely unfair and over-the-top, especially because Weigel had even defended Sonmez when she was under fire for posting a tweet about Kobe Bryant's 2003 rape allegations just hours after his tragic death in a helicopter crash in January 2020.

Sonmez responded by sharing screenshots of Del Real's tweets, adding: It's hard for me to understand why The Washington Post hasn't done anything about these tweets. It's almost as if Sonmez was trying to get him fired or suspended. It's almost as if Sonmez wasn't a sensitive hero, but a self-obsessed bully. Eventually—and shockingly—the *Post* seemed to catch on, firing her a little less than a week after this all began "for misconduct that includes insubordination, maligning your co-workers online and violating *The Post*'s standards on workplace collegiality and inclusivity."

Was firing Sonmez the best move? None of us really know the full story, so it's hard to say. It was likely wrong to fire her for her tweets unless there was more to it, but it was also wrong to dock Weigel a month's worth of pay over retweeting a stupid joke. When you think critically and logically, which is what we need more of from both individuals and institutions, it's hard to not come away with this outstanding truth: Twitter is a cesspool. Not much good comes out of it. It's not real life, so it's even more backward that we choose to enact real-life consequences on perceived tastelessness.

It's kind of wild to think how powerful a thumb has turned out to be, isn't it? But it's been happening for years.

In 2018, Roseanne Barr was, even in the words of (again) the *Washington Post*, "on the cusp of one of the great comebacks in television history." A full twenty years after the final episode of *Roseanne* aired, the premiere of a reboot of the sitcom had an astounding 27 million viewers. ABC renewed the series for another season, and things were going great—until Roseanne tweeted. On the evening of May 28, a Monday, Barr referred to former Barack Obama aide Valerie Jarrett as the child of the "Muslim Brotherhood & Planet of the Apes," a pretty big problem, considering that Jarrett is black.

The next day, Barr went *back* on Twitter to apologize:

> I apologize to Valerie Jarrett and to all Americans. I am truly sorry for making a bad joke about her politics and her looks. I should have known better. Forgive me-my joke was in bad taste.

Unfortunately for Barr, her apology worked out about as well as Arnold Schwarzenegger's 2017 "I take full responsibility for the hurt I have caused" when his nanny-baby news broke—and, like Arnold's, her mistake cost her millions. Her newly rebooted sitcom was canceled the same day. Later that year, ABC would launch a spin-off of the show, *The Conners*, featuring the rest of the family, but killing off Roseanne's character with an opioid overdose.

Sure, in Roseanne's case, her joke was obviously the worst kind: mean, stupid, hack, unhelpful, and, of course, racist. Plus, tweeting offensive things was a consistent Roseanne Barr problem. In fact, the *Post* article I cited earlier noted that network executives had repeatedly warned Roseanne about the possibility of her Twitter habit destroying the show. Hell, even her own son had hidden her Twitter password in a desperate attempt to make her stop. (Dopamine, man!) Despite her problem-tweeting being a pattern, though, it was still a single tweet that ultimately changed her life. One tweet, for which she was paid nothing, ultimately cost her everything. Of course, Roseanne's case was extreme—not only in terms of content, but also in terms of consequences. Still, she's far from the only comedian to run into a Twitter Problem.

Remember the #CancelColbert movement? In 2016, the official Twitter account for *The Colbert Report* attempted to satirize the Washington Redskins Original Americans Foundation, a group that argued the NFL team should keep its name for the sake of the wishes of Native Americans, with this joke: I am willing to show #Asian community I care by introducing the Ching-Chong Ding-Dong Foundation for Sensitivity to Orientals or Whatever.

The intention and message of the joke, of course, was actually progressive. As the *New Yorker* explained:

> *The joke, which originally aired on Wednesday's episode, is not particularly complicated: Daniel Snyder created a charitable organization for the benefit of a community and used a racial epithet for that same community in the organization's name—so here's an*

*absurd fictional extrapolation of Snyder's own logic. Everyone who hates both racism and Daniel Snyder laughs.*

*On Twitter, where words often slip free of their contexts, the unaccompanied punch line sparked a firestorm of outrage, which quickly escalated into a campaign demanding the show's cancellation. The hashtag #CancelColbert became one of Twitter's trending topics across the United States, and prompted Comedy Central to point out that the tweet in question, which was soon deleted, was posted by a corporate account that Colbert did not control.*

In the end, Stephen Colbert *wasn't* canceled, which is great, but the way the controversy unfolded really illustrates Twitter's problems. His scandal, like so many of them, treated one single tweet as if it stood alone—apart from intention, context, or anything else that Colbert had ever said or done. His whole body of work was condensed to a 140-character corporate tweet for which he wasn't even responsible. Colbert isn't the first, and certainly not the last, late-night host to get in trouble over Twitter posts. The year before, when Comedy Central announced that Trevor Noah would be the host of the new *Daily Show*, a BuzzFeed reporter dug up a few of Noah's tweets from 2009 through 2012, which featured offensive jokes such as:

> Almost bumped a Jewish kid crossing the road. He didn't look b4 crossing but I still would hav felt so bad in my german car!

> A hot white woman with ass is like a unicorn. Even if you do see one, you'll probably never get to ride it.

(I know. His spelling and grammar are *atrocious*!)

At the time, I wrote a piece for *National Review* defending Noah, pointing to a lot of the same things that I reiterate throughout this book: Noah is a comedian, comedians try out jokes, sometimes those jokes miss, but comedians need the freedom to try them out. Sometimes, they're also trying out a character. Comedians on Twitter regularly post tweets in the voice of someone who hates their spouse or their kids, or someone who has a drinking problem, or someone who's stupid or lazy. Sometimes, it's just a bit.

Noah, of course, wound up being just fine. Despite the controversy, he still took over *The Daily Show* and hosted it for seven years. I'm glad for that, but what makes me far less glad is how uncommon this sort of experience is, and the fact that it's usually much worse. In 2011, Gilbert Gottfried made a series of jokes about the tsunami and earthquake in Japan, the strongest recorded in the country's history, resulting in the deaths of roughly twenty thousand people: I just split up with my girlfriend, but like the Japanese say, "They'll be another one floating by any minute now," and Japan is really advanced. They don't go to the beach. The beach comes to them.

After facing backlash, Gottfried apologized, as you do these days, saying he simply was trying to make an "attempt at humor regarding the tragedy in Japan," and "meant no disrespect," which makes a lot of sense, considering how making attempts at humor—even about the difficult things in life—is usually part of a comedian's job.

Gottfried still lost his longtime deal as the voice of the Aflac duck. (It had been *so* longtime that I still heard "Aflac" in his voice when I wrote it just now.)

Twitter consequences aren't limited to professional comedians, either. Literally anyone, from any walk of life can be fired for dumb tweets, even if they were posted years prior as *teenagers*.

In 2021, a former Axios politics reporter, a twenty-seven-year-old black woman named Alexi McCammond, was slated to be the new editor in chief of *Teen Vogue*. But offensive tweets she'd posted ten years prior resurfaced, including Outdone by [an] Asian

#whatsnew, and now googling how to not wake up with swollen, asian eyes . . . and other tweets that used homophobic slurs.

Actually, the tweets didn't resurface so much as they re-resurfaced: McCammond had already apologized for the tweets and deleted them in 2019, acknowledging that they were "deeply insensitive." She apologized again when they were trudged back up to hurt her new job posting, saying, "I've apologized for my past racist and homophobic tweets and will reiterate that there's no excuse for perpetuating those awful stereotypes in any way."

McCammond was forced to resign a week before she even had a chance to start the job. But here's the thing: Of *course* McCammond posted some shithead comments in 2011. She was a teenager. When I was a teenager, I was smart. I was a valedictorian. There were ten of us, but still, my point is that I was good at school. I also said tons of stupid stuff, because no matter how smart I was, I was still a teen. Even the smartest of teens is still stupid. It comes with the territory. It's a rite of passage, in fact. No matter how smart I was, for example, I wasn't smart enough *not* to walk around my school with an Independent messenger bag that I'd gotten at PacSun with a giant Hot Topic patch that said CONFORMITY IS A SOCIAL DISEASE. I wasn't even a skater. Not only was I stupid enough to do that, but I was also stupid enough to actually wonder why no one wanted to talk to me. In short, I spoke and behaved as if I didn't have a fully formed brain. But guess what? That's because *I didn't have a fully formed brain.* Aside from my never-ending love for blink-182, I am thankfully not the same person as I was when I was a teenager. Which, by the way, means my dad *was* wrong when he told me I eventually wouldn't care about him not letting me go to the blink-182 concert in the seventh grade with Lauren and Amanda. He said it wouldn't be a big deal when I was older, but he was wrong. I *did* care. I *did* still think about it, and although I finally got to see them at Warped Tour in Atlantic City when I was thirty,

I was still upset that I'd missed my chance to see them play a show while Tom DeLonge was still in the band. Now that he apparently *is* back in the band and I have tickets to their reunion tour, I still can't help but think that the show won't be as good as it would have been back when they would be playing only Old Stuff (I know every word to every album from *Buddha* through *Take Off Your Pants and Jacket*, including *The Mark, Tom and Travis Show*, and if you don't believe me, just be thankful that you've never had to witness me proving it) because the New Stuff hadn't been written yet. But parental grievances are for another day. Or book.

Again, McCammond isn't the only one to suffer a reckoning over a tweet from childhood. In recent years, there have been several professional and student athletes who have been publicly brutalized over old tweets that they posted as teens.

In 2018, there was Josh Allen, a quarterback for the University of Wyoming who was a contender for the No. 1 pick in the NFL Draft. The night before the draft, a series of dumb, racist tweets—shocker—resurfaced, dating back to at least two years before Allen had even enrolled in college. Allen was the seventh overall pick in the 2018 NFL Draft, by the Buffalo Bills, but all the media could focus on were his tweets, not his prowess on the field. It wasn't about how far his arm could hurl a pigskin. It was about how far his thumbs had hurled his image.

That same year, twenty-four-year-old Josh Hader of the Milwaukee Brewers was playing in his first All-Star Game, and absolutely killing it (no, I won't get more detailed than that; google it, this isn't ESPN), when offensive tweets from his seventeen-year-old days started getting attention. Know what he got to do instead of brag about his All-Star Game appearance? He got to defend himself against his idiotic former self, repeatedly having to explain that he just wasn't the same guy he was when he was seventeen.

In 2019, there was twenty-one-year-old University of Oklahoma quarterback Kyler Murray, who won the Heisman Trophy on a Saturday, only to be the subject of countless headlines for something else all that weekend: the fact that he'd used the word "queer" as a homophobic slur in some tweets when he was fourteen and fifteen years old.

We could spend all day litigating the awfulness of some of their tweets, but that's been done for so many other days already. Plus, Hader's I hate gay people tweet isn't exactly a hot take worthy of weeks of examination. It's offensive. Full stop.

A conversation that interests me far more is this one: What is behind the desire to unearth old offensive tweets in the first place? What is it about seeing another person's success that motivates some people to delve into that person's past in search of something to destroy it? And, again, "resurface" is the operative word here. I'm not talking about a passive phenomenon. Unlike fuckboys, tweets don't have the ability to automatically pop back up as soon as they sense you're doing well without them. Someone has to go find them.

I could be wrong, but I have always thought there has to be some kind of sick, twisted jealousy at play. Perhaps these are people who have, unfortunately, not been able to reach the level of success they've always hoped for in their own lives. Perhaps these are people who think to themselves: *Hey, I may not be a Heisman winner or an All-Star pitcher, but I* can *take a Heisman winner or All-Star pitcher down!* If they can't get the power or prestige they seek through their own talents, they may decide to gain power by destroying someone who does have those talents.

The problem, of course, is that sabotaging the reputation of a Heisman winner doesn't make you any more of a Heisman winner yourself. It just makes you a contributor to a culture where a person's success can be jeopardized by something dumb he said when he was fourteen. That's not a world any of us should want to live in, especially considering how few people could ever hope to meet the standards of conduct that the Twitter Scold Monsters set for others. And that includes the Twitter Scold Monsters themselves! For example, Christine Davitt, a senior *Teen Vogue* staffer of mixed Irish and Filipino descent, posted a letter condemning McCammond's tweets—only for news to break that Davitt had used the n-word in tweets of her own from over ten years prior.

So, why, then, would Davitt go after McCammond, knowing that she herself has not been perfect? Why would anyone go after anyone, knowing what might happen as a result? Is your desire to virtue-signal really more important than another person's job? A 2019 piece in the *Atlantic* by Jonathan Haidt, a social psychologist and professor of ethical leadership at New York University Stern School of Business, and Tobias Rose-Stockwell, a technology ethics writer, theorized that part of the problem might be "the way social media turns so much communication into a public performance" rather than the "two-way street" that it is elsewhere.

In the piece, they discuss psychologist Mark Leary's claim that so-called *self*-esteem is actually based on where we think we rank socially according to our internal "sociometer"—which is Leary's term for our brains' ongoing evaluation of what *other* people think of us.

Social media, Haidt and Rose-Stockwell argue, can serve as a sort of public sociometer, one that's scored based on the popularity of our accounts and posts. Because it's public, we're motivated to score high not only to gain that approval from others for ourselves, but also in order to display that approval from others *for* others when they look at our accounts.

The problem? Unlike in typical social interactions, where repeated displays of anger would probably be considered exhausting, social media metrics seem to *reward* outrage. Haidt and Rose-Stockwell cite two studies to back this up: one study conducted by NYU researchers in 2017, which found that each "moral-emotional" word found in a tweet made it 20 percent more likely to be retweeted, and a Pew Research Center study from the same year, which found that Facebook posts displaying "indignant disagreement" got double the engagement of other posts. (Which low key may explain why your otherwise-darling aunt insists on acting like *that* on the platform.)

Haidt and Rose-Stockwell invoke a phrase coined by philosophers Justin Tosi and Brandon Warmke to describe what happens on social media under these approval-seeking conditions: "moral grandstanding," or public displays of moral virtue with the aim of making yourself look better.

In the "competition to gain approval of the audience," Haidt and Rose-Stockwell explain, "nuance and truth" are thrown out the window in favor of overwrought emotional displays.

"Grandstanders scrutinize every word spoken by their opponents—and sometimes even their friends—for the potential to evoke public outrage," Haidt and Rose-Stockwell explain. "Context collapses. The speaker's intent is ignored."

People don't really consider the implications of what social-media-shaming someone might have for that person, because, well, they don't have to. They don't know the person being piled on, and they will almost certainly never have to face him or her, either. Under these circumstances, the under-fire person becomes more of an abstract opportunity for obtaining social capital than a person at all. Plus, it's so tantalizingly easy. It doesn't require any sort of original thought: Find someone who posted something offensive and quote-tweet it with "Wow" or "Gross," or find someone else complaining about an offensive tweet, and quote-tweet *that* with something like "This" or "Seriously." That's all you have to do in order to signify yourself as One of the Good Ones on the Good Side, and it costs you absolutely nothing. You don't have to explain to the person why you contributed to his or her cancellation in order to gain points for yourself. Hell, you don't even have to get off the toilet.

Some people would argue that calling someone out for past insensitive comments isn't the same as calling for that person to lose everything. Pssh.

As Megan McArdle pointed out in a 2022 piece in the *Washington Post*, that view is more than just a little out of touch with reality. It opened like this:

> *At this late date, it seems almost unnecessary to point out that if you publicly accuse someone of racism, sexism or other similar wrongs, you are effectively calling for that person to be fired, or at the very least, to suffer some kind of workplace discipline. Yet apparently someone needs to restate the obvious.*

McArdle was writing in response to the controversy surrounding a poorly worded (what else!) tweet from legal scholar Ilya Shapiro, who was then slated to become the executive director of the Center for the Constitution at Georgetown Law, about Joe Biden's campaign promise to nominate a black woman to fill the Supreme Court vacancy created by the retirement of Justice Stephen Breyer. In the tweet, Shapiro insisted that the best pick for the job would have been Sri Srinivasan, an Indian-born judge sitting on the US Court of Appeals, and that it was unfortunate Srinivasan didn't "fit into latest intersectionality hierarchy so we'll get a lesser black woman."

McArdle called Shapiro's tweet "offensively worded," and I agree. In general, I would say that if you are about to do a tweet, and see that you've included the phrase "lesser black woman," you might want to rework it. But McArdle's focus, rightly, was on the way the controversy ensued and the predictable consequences.

The tweet didn't get too much attention on its own, offensively worded though it was, because presenting identity politics as being at odds with meritocracy is hardly an unusual or extraordinary take.

It was only after Mark Joseph Stern of Slate shared screenshots of the tweets, along with the comment that he had felt compelled to do so because he felt "an obligation to condemn his overt and nauseating racism" and was "deeply ashamed of [his] alma mater," Georgetown University Law Center, that the post got any attention.

The consequences of this attention were, like I said, entirely predictable. Shapiro apologized, acknowledged "inartful" wording, and deleted the tweet—but people were still upset. "Nonetheless, it was obvious to everyone that Shapiro's job was on the line," McArdle wrote. "Except, apparently, to Stern, who insists that he never intended to get Shapiro fired."

McArdle pointed out how common it was to see Internet scandals play out exactly like this: An ever-growing mob decries the offensiveness of a comment, the person who made the comment gets fired over it, and the mob insists that, although they may have been upset, it's not like they intended for anyone to get *fired* or anything! So why, she asked, do people keep participating in the Internet pile-ons that so often lead to firings—and why do they result in firings so often—if no one seems to want to see firings at all? She rejected the idea that anyone could actually be surprised to see a firing after "so many examples" of them, and offered this explanation instead:

> *More likely, many personally think firing is too extreme, while nonetheless feeling impelled toward the inevitable outcome. Initiators want to call out bigotry, those who pile on must comment on the issue of the day, and employers cannot face days and weeks of scandal.*
>
> *Enthusiasts for these mass shamings talk about holding people accountable for the intangible harms their words cause. Yet they fail to take responsibility for the very tangible harms they inflict when they launch the first fiery salvo, or furiously click "retweet."*

> *Underneath Shapiro's appalling word choice lay a vital moral and political question: Is it legitimate to rectify past discrimination with current discrimination? I'd argue that it is, not because today's White males deserve to suffer for the sins of their forebears, but because demographic representation enhances democratic legitimacy.*
>
> *As we have done so many times before, we turned one of the most sensitive, complex and important issues of our day into a binary referendum on one person: Ilya Shapiro, racist or not?*

Shapiro was quietly reinstated in June 2022, after a nearly six-month investigation into the incident, but ultimately decided to resign, writing in the *Wall Street Journal*: "Dean William Treanor cleared me on the technicality that I wasn't an employee when I tweeted, but the IDEAA implicitly repealed Georgetown's Speech and Expression Policy and set me up for discipline the next time I transgress progressive orthodoxy. Instead of participating in that slow-motion firing, I'm resigning."

McArdle's take stands out to me because it does the opposite of what everyone seems to do to stand out: It shows nuance. McArdle didn't write her piece because she thought Shapiro's tweet was great. Not only does she call his word choice "appalling," but she also says that she doesn't even agree with his underlying point. Yet she is still able to take a step back and really see the cultural consequences of our tendency to reject real conversations about complicated issues in favor of cheap, surface-level ones about whether someone's single tweet makes him Bad or Not Bad.

If the only two options in a Twitter controversy are to condemn or defend a person—like you're voting them on or off of an island without Jeff Probst to guide you—then we're missing out on a lot of other conversations we could (and should) be having. This sort of binary approach inherently prohibits you from thinking critically and exploring all angles of an issue; it allows for only two possible takes at the expense of all others. What's more, it's not just the fact that these issues are complex and therefore deserve complex consideration, because there's also this: *People* are complex, too. Every single human being is an enigma, and far too mosaic to ever be entirely encapsulated by a single tweet, joke, or comment.

Ricky Gervais made a similar point as it applies to comedy: A single joke does not tell you everything you need to know about the person telling it, and it's absurd how often people behave as if that isn't the case. Gervais made the point in an interview with the *New York Times* in May 2022 after the release of his Netflix special *SuperNature*, in which he made jokes about everything from a dead baby to Muslims to God to pedophilia to—the topic which would create massive backlash—trans people. Or, as he referred to them at one point in the special: "[T]he new women. They're great, aren't they? The new ones we've been seeing lately. The ones with beards and cocks."

He told the *Times*:

> *I think that's the mistake people make: They think that every joke is a window to the comedian's soul—because I wrote it and performed it under my own name, that that's really me. And that's just not true. I'll flip a joke halfway through and change my stance to make the joke better.*

The backlash Gervais faced over his jokes was extremely intense—including declarations on social media that his jokes would kill trans people, and vows among some to cancel their Netflix subscriptions over it. This is despite the fact that, later in the special, Gervais made sure to clarify:

> *Full disclosure: In real life of course I support trans rights. I support all human rights, and trans rights are human rights. Live your best life. Use your preferred pronouns. Be the gender that you feel that you are.*

Gervais did follow that up with "But meet me halfway, ladies: Lose the cock. That's all I'm saying." But the point still applies: As humans, we *are* complex creatures. We have all said and done bad things, and saying something bad or unsavory doesn't automatically signify you're a bad, worthless person overall. No one's entire essence can ever be summed up in a few hundred characters—and certainly not by a joke or a tweet we made when we were children.

It's also important to remember that Twitter, again, is not real life. There's data that backs this up. According to a recent Pew Research Center study, only one in five Americans claim they even use Twitter at all. What's more, the top 25 percent of users, in terms of the number of tweets posted, are responsible for a whopping 97 percent of tweets. Therefore, a Twitter Consensus on an issue just isn't the same as a real-life consensus on an issue. Unfortunately, the twenty-four-hour news cycle also plays a part in fueling this outrage. Or, as Gervais put it:

> *Twenty years ago, if you complained about something, you get out a pen and paper and go, "Dear BBC." Now they can fire off a tweet, and the fucking press pick up the tweet. They'll say, "People are mad, said 69341" . . . most people aren't mad. Most people don't know about it.*

He's right. All too often, it *feels* like everyone agrees you can't joke about something—just because most of Twitter says you can't joke about something. And let's keep in mind that, as of this writing,

Twitter has roughly 330 million active users, while the world has roughly 8 billion people living in it. Twitter opinions are hardly the majority. Most of the world is living outside the cesspool.

To Gervais's point: A common form of news headline these days is the People Are Mad story. It's something that gets presented as a massive, explosive controversy, even though the only evidence of that is a small handful of posts from a platform that a small handful of people use. I know this not only because I've seen it, but also because I've done it. I have, admittedly, years ago, published "People Are Mad" pieces using Twitter as evidence without investigating further. I did it because other people were doing it, or because I was busy and needed to fill an article quota and get back to my other (or, depending on the year, multiple other) jobs. It was stupid, it was lazy, and I regret it. There have been times that I got dragged for it, and deservedly so.

I bring all that up not because I want to get shit on again (as far as kinks go, that isn't really my vibe) but because I want to make it so, so clear how easy it is to fall for outrage bait. It's so simple, and so seemingly low-stakes, that it's an easy trap to fall into—even if you *do* fancy yourself as someone who stands against all of it on principle.

It's important for all of us to do our best to resist the temptation to pile on, no matter who is under fire. If you're a conservative, for example, it's not enough to just defend the likes of Donald Trump–loving Roseanne Barr—you should also have defended, say, Sarah Jeong.

In 2018, Jeong had just joined the editorial board of the *New York Times* when some tweets she had written about white people in 2014—such as Dumbass fucking white people marking up the internet with their opinions like dogs pissing on fire hydrants and It's kind of sick how much joy I get out of being cruel to old white men—came to light. Jeong insisted that her tweets had been satirical and in response to people who were attacking her, but conservative outlets ran pieces on her "racism" anyway. Now, to be fair, many conservatives used the incident to point out that, if the shoe had been on the other foot, one of them would not have been given the same grace as she was given. Many pointed to Kevin Williamson, who had just been fired from the *Atlantic* over some past offensive comments about women who have abortions.

I do think you can make a pretty good case for a double standard. And I'm not just saying that because I know and respect Kevin, a former colleague of mine whom I consider to be an absolutely brilliant writer, even if I do disagree with him on several issues, abortion being one of them. But I just don't think that's the most important discussion for us to have when this sort of thing happens. It's far more important to unequivocally stand up against a mob that seeks to destroy someone's career over some tweets, as if some tweets are a bigger indication of a person's career worthiness, above and beyond anything and everything else they have ever done to get there.

To Kevin's credit, he himself published a piece in *National Review* titled "The *Times* Can Hire and Fire Whomever It Likes," imploring fellow conservatives not to use his name to demand Jeong's firing. I was glad to see that Jeong didn't get fired, and that I actually didn't see too many, if any, prominent conservatives calling for her to have been. In other instances, though, that hasn't been the case. In Williamson's same *National Review* piece, for example, he said he also hoped Marvel would not fire filmmaker James Gunn over *his* old tweets; Gunn was not as lucky.

Between seven and eight years earlier, Gunn had posted some pretty disgustingly offensive jokes about pedophilia and rape, which right-wing bloggers starting drawing attention to in 2018. In response to the controversy, Gunn apologized, explaining that he had "developed as a person," and so had "[his] work and [his] humor."

I understand, at least intellectually, how appealing partisan fighting can be among people who have a party allegiance, but it's a temptation that should be resisted at all costs. Even if conservatives do bear the brunt of cancel culture, bearing the brunt of something shouldn't make you want others to suffer, too. Rather, it should motivate you to want things to fundamentally change. For everyone. And the best way to do that is what your mother has been telling you to do from childhood: Lead by example. Are there some people on Twitter who are genuinely bad people? Yeah, bro. But an "offensive" post in itself is not evidence—let alone a closed case—that a person is an incorrigible wretch. It's time we all recognize the limits and pitfalls of social media and associated outrage and start treating people as exactly what they are: complex and deserving of the kind of consideration that acknowledges that reality.

CHAPTER 14

# COMEDY IS MY RELIGION

I used to be super, super Catholic.

And by "super, super," I do mean "super, super." I was an altar server. I not only went to Confession, but I also told the truth there—even after I reached an age where my sins started being the sorts of things that were weird to talk to an old man about, but I did so anyway because the only thing scarier was my certainty that I'd burn in Hell for eternity if I didn't. I voluntarily did bread-and-water fasts on Wednesdays in the fourth grade for the sake of the lost souls in Purgatory. Everything.

I fell away from it completely, probably sometime in college. These days my views on a higher power are basically that I have no idea if there is one, and I doubt that I'll ever know, but hope that I will. Sincerely, I do hope that I can somehow figure it out before I die, so that dying won't be so scary. If not, I guess there's always opiates, which seem to work pretty well. Any time I hear people commend someone for being strong and "in good spirits" despite battling a horrible medical situation, I always wonder if it's *really* that person's innate courage, or more so their Dilaudid drip.

I often find myself feeling jealous (a sin, I know) of religious and spiritual people, because of how much I wish that I could still be one myself. It would be so much easier to believe that there was an all-powerful being looking out for me all the time, or that the trajectory of my future would be something other than continuing to get older, eventually to the point that I'm completely unfuckable, and then dying—and it's over, just like it was before I was born.

It would be nice, I think, but I just can't find a way to convince myself. Still, there is another area in life that provides perspective, facilitates healing, and brings people together: jokes. So, for now, the closest thing that I have to any sort of religion is comedy. Laughing and making people laugh is easily my favorite thing about being alive. The idea of comedy being my religion might sound kind of crazy, and I totally understand that it is far from a perfect replacement, especially since it doesn't offer me the promise of eternal life the way that actual religions do. At the same time, though, there are a lot of parallels between the two if you really think about it.

Science shows that laughter has some of the same positive impacts on our brains as some religious services do. Formal religious rituals are associated with an increase of dopamine, serotonin, and oxytocin in people's brains, making them feel happy—and research shows that laughing produces those effects, too.

An obvious parallel between comedy and religion—and one directly related to that shared brain-chemical impact—would be the one I've discussed throughout this book already: the ability to provide healing from pain. It's no secret, after all, that people turn to religion for comfort in times of distress. Hell, even people like me (lost, godless heathens in search of concrete beliefs) can sometimes find ourselves compulsively praying, almost like a knee-jerk reaction, when we find ourselves feeling like we might be really, really, really fucked.

There's science behind this, too. An epidemiologist at Rush University Medical Center in Chicago, Lynda H. Powell, reviewed approximately 150 papers studying the connection between health and religious faith. In her research, she found that religion gave people emotional comfort in times of sickness, which isn't surprising, and seems almost too obvious a conclusion to have wasted resources reaching it. But there was more: Powell found that, although faith wasn't associated with outcomes such as the faster resolution of an acute illness or slowing the progression of cancer, it *was* associated with a 25 percent lower mortality rate—*even after* adjusting for other variables such as health and lifestyle choices. Pretty amazing, right? An article in *Newsweek* discussing her findings also noted that studies of brain scans determined that meditation/prayer could have similar impacts, lowering people's heart rates and blood pressure, as well as boosting their immunity.

The power of comedy in terms of coping emotionally with difficult or even traumatic situations is something that I've discussed repeatedly throughout this book—the way that laughing at something devastating can take away its power, those repatriated Vietnam War prisoners who claimed making jokes about their captivity was even more helpful than religion in getting them through it, and a million more examples. Comedy has been a huge part of my own support system, one that I've desperately needed through the toughest times. Although I may have spent my entire childhood, ages consciousness to eighteen, terrified that I was vulnerable to an evil monster in the form of the devil (yes, even though my mom had put Saint Benedict medals above all the doors of our home to keep him out), I also believed that I had an all-powerful support system in the form of an all-powerful God. Of course, it's impossible to replace that. I sometimes wonder if the hole left by my loss of belief has, at times, led me to be too needy in some of my relationships with some humans. Still, I can't imagine how much more difficult things would have been—both within myself and with others—if I hadn't discovered comedy's capacity to provide a light in the darkness. If you're this far along in the book, though, then you've basically already become an expert on all of that by reading my powerful syntax sprinkled with scintillating anecdotes. I will add, though, that science also says that—just like with religion—the healing powers of comedy go beyond the emotional. Like religion, comedy and laughter can make a difference in terms of physical healing, too. Psychologist Dr. Pamela B. Rutledge, director of the Media Psychology Research Center in California, explains that the release of dopamine, serotonin, oxytocin, and endorphins that happens in response to humor "decreases stress, diminishes pain and in the process strengthens the immune system."

The physical benefits of humor are well documented. In addition to releasing those feel-good brain chemicals, laughing also improves circulation, reduces blood pressure, and even increases the body's production of T-cells and antibodies that help ward off infection. (Unfortunately, also like with religion, I was unable to find any empirical evidence that jokes were able to stop the spread of a person's cancer.)

Of course, a religious faith ostensibly provides people with guidance toward good behavior, encouraging them to be better versions of themselves. I mean, it may not have worked for that Ted Haggard guy, who used his position as an evangelical celebrity pastor to preach the evils of homosexuality even though he was reportedly smoking meth and banging a male prostitute in his free time. (To be clear, the only thing I have a problem with there is the hypocrisy and the deceit and the spreading of homophobia. If not for that, I'd say: *Do you, Ted.* The rest of it, although not my thing, is all victimless stuff.) All Ted Haggards aside, *some* facets of some religious doctrines could make society a better place. I mean, the ones that tell you things like "Don't have sex until you're married" and "Never, ever have gay love" are not things I will ever be able to get behind, because I think they're damaging and wrong. But other teachings, like telling people that we should love and care for each other? I vibe with that; I think that's great. Especially if I'm not in a mood and no one is pissing me off. Still, regardless of your view on any specific teaching itself, the aim of them overall is clear: to give people guidance for molding their behavior.

Doesn't comedy kind of do that, too? Since forever (whatever, this isn't a history book), comedians have been using their platforms to call out behavior that they see as socially or morally unacceptable, or even just as annoying, by mocking it with their jokes. When Elayne Boosler famously joked that men "want you to scream 'You're the best!' while swearing you've never done this with anyone before," she wasn't just looking for a laugh; she was also calling out straight male hypocrisy on sexuality and the unfair sexual double standard between men and women. When Lenny Bruce joked, "Never trust a preacher with more than two suits," he was chastising religious leaders who exploit faith to enrich themselves. Or, on a (far) smaller scale, when Jim Gaffigan joked, "I can't believe we're still giving clothing as a gift. 'Cause whenever you get clothing as a present, you always open it up and you think, 'Not even close.' And the person that gives it is always like, 'You can take it back if you don't like it.' 'That's all right. I'll just throw it out.' Don't give me an errand," he was aiming to give all those relatives who haven't seen you often enough to have any idea how big you are helpful gift-giving advice. Just as religion seeks to shape human behavior with its teachings, so does comedy with its jokes.

Another reason people turn to religion is to feel a sense of meaning in life, and I totally get that. Again, I wish I had it in me to believe that I am only *temporarily* this aging bag of bones and blood held together by a skin sac, that I will someday be a celestial being looking down on my loved ones from Heaven. And, more importantly, able to read all the RIP social media posts about how I was so brilliant and funny and will be sorely missed. (Like, just tell me while I'm alive, people.) Instead of having spent months crying over that guy who broke up with me in front of my father at Coney Island six months after my mom died, instead of wasting away and chasing him and thinking I had destroyed the only hope I'd ever have of dying any way but alone (I'll never forget explaining that fear to my friend, who contested it by joking: "No way. Murder-suicide"), I would have loved to have been able to tell myself, as I was sobbing and snotting both in my bed and in bars around the city (thanks for being there, Ben Kissel), that *everything happens for a reason.*

"Everything happens for a reason." Man, people *love* that one, don't they? It's easy, of course, to see why: Not only is it a great way to kind of absolve you, yourself, as the cause of any of the horrible things in your life (it's what the Universe wanted!), but it also adds a layer of meaning to absolutely everything that happens, bad or good. It makes the horrible tolerable; it makes the mundane meaningful.

People are *desperate* to find meaning in their lives. I mean, there are even people who will purposely burn the poison of a frog into their skin—making them vomit and sweat as their faces swell up several times their normal size—in an attempt to cleanse the physical and emotional toxins from their bodies and brains before smoking a psychedelic toad to zap their ego in an indigenous ceremony administered by a shaman while sitting on a blanket on the floor of a friend's New York City apartment.

Comedy can offer a sense of meaning, too. It gives you that Zen perspective you can't get many other ways. Without a religion, it's really all that I have—especially because I have never done that weird frog/toad thing. Clearly, only total psychos would do that.

For example: That breakup may have been devastating, but it did give me a great story to joke about on *The Greg Gutfeld Show* one weekend years later. I had mentioned offhand on the show that a guy had once broken up with me in front of my father, and both Greg and Tyrus immediately demanded I tell the rest of the story: "My dad went to go get more drinks, and he broke up with me while my dad was getting more margaritas, and he came back, and I was just sitting there crying. We were at Coney Island. Just when you thought it couldn't get worse, it just did. And then he hung out with us the whole rest of the day, and he came back with us on the train. It was the weirdest family outing ever. And if he's watching, I'm over it."

Then later interjecting, "I would just like to add that this was a first-degree breakup. He came to Coney Island knowing he was going to break up with me. It's not like I did something while my dad was getting drinks. He came there to dump me.

"Our drinks were done, my dad went to get the second round, and then he just broke up with me, and my dad came back to a shitstorm."

"*He sat next to me on the Ferris wheel afterward!*" I screamed.

The story wound up being, I think, one of the best moments on the show. It was totally unexpected, as was how much worse it would continue to get the more details I offered. The audience was laughing and applauding, and at one point, Greg said, "I'm just realizing that this never happens on *Special Report*!"

That breakup, especially considering that it did happen just six months after my mom died (and three months after my grandma died), and that he was the guy who had been there with me through it all, was one of the most brutal, awful times in my life. Now, I *am* glad it happened, and for a lot of reasons. One of them, of course, is that Coney Island Guy and I now get along great as friends, and I am now married to someone who is a far better match for me than he ever would have been. But that isn't the only reason. That experience, after all, also gave me an incredible, captivating story to tell. I have often said I am grateful to all of the POS men

I dated in my twenties for the content they have provided me, and what's more, I actually mean that. That stuff was excruciating to go through, sure, but I ultimately decided to find things that were funny about it, to turn it into material to make myself, and then others, laugh. Throughout this book, it's obvious how often I do this: Whenever I'm going through something tough, I find the comedy in it, and find that doing so makes me feel better. The thing is, though, it's about more than just that little therapeutic boost that the laughter supplies; it's also the meaning it can provide to those difficult experiences. I get to say: *Yes, this is tough, but look at what I can create from it . . . and how much power I can feel by doing so.*

I'm not the only one who thinks of it this way, either. In fact, Viktor Frankl—an Austrian psychiatrist, philosopher, author, neurologist, and *Holocaust survivor* who founded logotherapy, which describes looking for meaning to be the most motivational force in human life—wrote about this exact thing in his book, aptly titled *Man's Search for Meaning*:

> *Everything can be taken from a man but one thing: the last of the human freedoms—to choose one's attitude in any given set of circumstances, to choose one's own way.*
>
> *It is this spiritual freedom—which cannot be taken away—that makes life meaningful and purposeful. An active life serves the purpose of giving man the opportunity to realize values in creative work, while a passive life of enjoyment affords him the opportunity to obtain fulfillment in experiencing beauty, art, or nature. But there is also purpose in that life which is almost barren of both creation and enjoyment and which admits of but one possibility of high moral behavior: namely, in man's attitude to his existence, an existence restricted by external forces.*

There is, of course, immense value in creative work—and not just for the people who get paid for it like I do. There's power in deciding to look at something that you may feel could destroy you, and to decide to create from it instead. Plus, going back to something like that Coney Island breakup story, for example, I have heard from people that my tendency to (some would say *over*) share horrifying or humiliating stories about my life has had some kind of impact on them. They've told me that they've found some of it relatable in some way, or that it encouraged them to talk about some of the issues in their own lives that they wouldn't have otherwise.

Which brings me to the next thing that religion and comedy have in common: As anyone who has a relative who fills up your entire Facebook feed with photos of potlucks and mission trips can tell you, religion brings people together—and it's not hard to see how comedy does the same thing.

For example, I can't be the only one who has like five or ten friends with whom I would have totally lost touch if not for us sending each other funny memes back and forth. People can move, and people can change, but there will be memes that you think they'll find funny forever. There's also nothing better than joking around and laughing with a group of friends; I'm so obsessed with it that I had to make it my job. Plus, laughing with others also bonds you with them in the biological sense—again, laughing releases oxytocin, which is literally called the "bonding chemical" or the "empathy hormone" because of the way it can create feelings of closeness.

Honestly, if you really think about it, a comedy club isn't all that different from a worship service. It's a group of people gathered together to hear someone talk about life. Like the people in the pews of a church, the people at the tables of a comedy club are expected to pay full attention to the person with the microphone—to turn off their phones and to listen without interrupting, responding only with smiles, laughter, and relevant head movements.

There are, of course, some differences. For example, there are usually no babies in a comedy club, and that's a good thing. In a church, though, it is a *great* thing that there *are* babies. At least in my opinion. Back when I was younger and went to church with my family, I remember thinking that nothing made the whole thing entertaining quite like getting to watch some parents doing their best to wrangle a totally out-of-control rugrat the entire time. Another difference: At a comedy club, people can usually eat while they watch. In church, you can't really do that—unless you are a baby, in which case you may eat the Cheerios that you have dropped onto the floor.

Pastors and comics may be different—I say *may* be because I haven't met every single one—but their roles in that moment of preaching or performing are actually quite similar—even if we aren't talking about the pastors who try to spice up their sermons with jokes. Both stand-up comics and pastors have control of the room, with the expectation that what they're going to tell you is something that's worth hearing. Although, of course, at least in some religions, the church leader has the added draw of offering you the whole Not Going to Hell benefit of attendance, too.

I've never been a pastor (could you *imagine*?), but I used to perform stand-up comedy every single night. I don't really do it anymore, and I am not sure when I will again. The last time I did it was in 2021, when I decided to get back onstage for the first time in a year and a half for a set at a Nashville taping of *Gutfeld!* that would air on television in front of millions of people, with only one night to prepare. I did it because I am insane, and "no" is not the word that I'm best at, but also because stand-up has proven to be impossibly difficult for me to quit. So, although I'm not sure when I will do it again, I am pretty sure that I probably will. It truly is a disease. As brutal as stand-up and the associated lifestyle can be, it's also a high like nothing else. Bombing makes you (or at least me) feel at least a little suicidal, but the flip side is that there's nothing in the world like having an amazing set.

I don't do it so much anymore for a few reasons. The main reason, honestly, is that I can no longer devote the time to it that I would need to do it well. I've started focusing on other things, like my TV job and this book, and I can't do both at the same time. Kind of like that priest at one of my old churches who quit so he could get married to his girlfriend. Except he pretty much definitely can't ever go back, even if he does miss it as much as I miss stand-up. The only thing harder than being a news personality *and* a stand-up comic is being a husband *and* giving Communion in a Catholic church.

Plus, to be honest, I always worry that there would be a chance I'd say something in the middle of a riff that would turn the audience—not in the room, but The Audience in the greater sense, especially with people having camera phones on them—against me. Which brings me to something that religions seem to do better than comedy: forgiveness.

Most religions make it a point to contain some path to forgiveness for your wrongs. In Islam, you can seek redemption through making amends to both Allah and the person you've wronged, praying, and doing acts of charity. In Judaism, you have to sincerely apologize to the person or people you've wronged. (Kind of bad news for murderers, but otherwise a pretty good deal.) In Catholicism, regardless of the sin, you have to feel sorry for what you did, confess it to a priest, and do whatever he (I almost habitually wrote "or she") tells you to do as penance, usually something like a few Hail Marys, and then you're saved from eternal suffering in Hell.

When I went to a Lutheran high school, I learned that those guys had it way easier. As long as you totally accept Christ as your savior, you don't have to worry about going to Hell. (No telling an old man about what you can't seem to stop doing with the showerhead necessary!)

Actually, even one of the most fire-and-brimstone religious texts out there—the Old Testament—has a more lenient standard of punishment than what our culture sometimes levies for making an errant joke nowadays. Leviticus (easily one of the least chill books in the Bible) says, "If anyone injures his neighbor, as he has done it shall be done to him, fracture for fracture, eye for eye, tooth for tooth; whatever injury he has given a person shall be given to him." Here's the thing: If you were to ask people to explain the worst, most traumatic life experiences that they've ever had to endure, I'd bet that absolutely *none* of them would put having to hear a bad or offensive joke anywhere near the top of their list. If you've made it this far in this book, however, you already know that having *told* a bad or offensive joke might make that list for at least some people.

As harsh as that Leviticus standard may have been (especially when weighed against the Gospel of Matthew, which updated this advice to say that you should actually "turn the other cheek" if someone hits you), it's still not as severe as what our culture does in terms of comedy and speech. Even the Leviticus framework would say that the furthest someone should go to pay someone back for his nasty joke would be to fire a nasty joke back at him.

To be clear: I'd be cool with a much lower forgiveness requirement for comedy than what's demonstrated in some religions. After the Romans crucified Jesus Christ, the Bible said that Jesus's dying words were, "Forgive them, Father, for they know not what they do"? I mean, good on Jesus, but there is *no way* that I'd be having that kind of attitude about people who had just hung me to a cross.

I definitely get why Judas was, at least according to what I've googled about him, damned to an eternity in Hell. Like . . . Judas, bro, you're really going to *kiss* Jesus as a way to show to his future murderers that he's the one they should kill? That is *low*. I've had a bad enough time dealing with someone who kisses me acting weird about it the next day.

Thankfully, there aren't any comedians out there who are (to my knowledge) involved in crucifixions, so the potatoes are considerably smaller. Given that, couldn't we do at least just a little better when it comes to forgiveness in terms of comedy?

In 2019, Sarah Silverman said that she had recently been fired from a movie (at 11 p.m. the night before the gig was supposed to start, no less) over a 2007 episode of *The Sarah Silverman Show* in which she wore (you guessed it, baby!) blackface. The sketch's intention, as was the intention with so many if not all of these recent examples, was not to create harm. It was part of a sketch where she satirically used blackface in exploring whether it was more difficult to be Jewish or black.

In 2019, she discussed the firing on a podcast, and said, "It was so disheartening, it just made me real, real sad because I've kind of devoted my life to making it right."

Now, there's plenty of stuff, doctrine-wise, that I disagree with when it comes to religion and forgiveness. Like that thing in the Catholic Church where, if you masturbate, and then head out to get donuts, but then die in a car crash on the way there without going to confession, then that means you're going to Hell. I mean, I just don't think that masturbation is something you should even need forgiveness for in the first place—and I would think it was just so awful that you died instead of getting your donut.

At least in the Catholic Church, though, intention actually matters. Grave sins are only considered grave sins if you knew that they were grave sins, but still went ahead and did it anyway. Plus, no religion, to my knowledge, takes the attitude that you *can't* be forgiven for mistakes, *even* if you *spend the rest of your life* trying to make up for it.

Sarah Silverman, for example, wasn't exactly "canceled." She's still famous and rich despite the blackface scandal; she only lost one movie job that we know about. I also don't know if it's true that she, in her words, "kind of devoted [her] life to making it right," because I'm not really in a position to follow her life closely enough to make that kind of determination. Still, it seems totally unfair for a comedian to lose any job for a joke she'd made twelve years ago, even after she'd expressed regret over it several times in between, including during a *GQ* interview the year before, when she said, "I don't stand by the blackface sketch. I'm horrified by it, and I can't erase it. I can only be changed by it and move on."

Again, as much as I may disagree with what most religions consider "sins," I appreciate the way that they tend to offer a path to redemption. Interestingly enough, Silverman called for exactly this in a 2020 episode of *The Sarah Silverman Podcast*:

> *In this cancel culture, and we all know what I'm talking about, whether you think there is one or there isn't one or where you stand on it, and there's a lot of gray matter there, but without a path to redemption, when you take someone, you found a tweet they wrote seven years ago or a thing that they said, and you expose it and you say, this person should be no more, banish them forever.*
>
> *If we don't give these people a path to redemption, then they're going to go where they are accepted. . . . I think there should be some kind of path. Do we want people to be changed? Or do we want them to stay the same to freeze in a moment we found on the internet from twelve years ago?*

Here Silverman is discussing a friend of hers who had previously been a leader in a hate group because he was, as she saw it, just looking for love and acceptance somewhere. It's an excellent point: If we really want people to be on the side of love and acceptance, we have to be willing to love and accept them even after they've made mistakes. It's not only a way to be kind to someone, but also an incentive for that person to genuinely change. Refusing to do so, on the other hand, would likely incentivize them to dig their heels in even further, because why wouldn't they? Just as the ability to be divinely forgiven is certainly something that attracts people to religion, the ability to be forgiven by a human whom you've hurt could certainly attract a willingness to listen and learn.

Also, forgiving people is good for you. No, I don't mean in that woo-woo way that you might read in the caption of an influencer's Instagram photo of her Pretend Meditating outside of some temple in Tulum—saying something about how she's *releasing* anger, when the obvious real motivation for the photo is that she's *releasing* the top two-thirds of her jugs from her Gymshark sports bra. I mean this biologically. According to Johns Hopkins University, "[s]tudies have found that the act of forgiveness can reap huge rewards for your health, lowering the risk of heart attack; improving cholesterol levels and sleep; and reducing pain, blood pressure, and levels of anxiety, depression and stress."

Plus, there's this: Without forgiveness, comedy just can't exist. Mistakes are inevitable when it comes to comedy. It's going to be more common when it comes to situations where you're joking about a tough subject, but tough subjects are the ones that need jokes the most.

Forgiveness in comedy has to be crucial, because comedy is crucial. Not just for me—remember, people who have been through really, *really* tough stuff, like the Holocaust, or being a war prisoner, happen to agree with me on this.

As for me? It's the closest thing I have to a religion . . . so please don't destroy it. It may not offer me any promise of life after death, but honestly, I'd at least like to be able to keep laughing in the meantime, until I inevitably crash my car on the way to the donut shop.

# PSA: YOU ALSO HAVE THE RIGHT NOT TO SPEAK

I talk a lot in this book about freedom of speech, but I also just want to remind everyone that you also have the freedom to *not* speak—a right that you should always take advantage of if you ever find yourself in an encounter with the police.

Here is some advice for handling encounters with police, as per the American Civil Liberties Union:

### I've been stopped by the police in public

YOUR RIGHTS: You have the right to remain silent. For example, you do not have to answer any questions about where you are going, where you are traveling from, what you are doing, or where you live. If you wish to exercise your right to remain silent, say so out loud. (In some states, you may be required to provide your name if asked to identify yourself, and an officer may arrest you for refusing to do so.)

You do not have to consent to a search of yourself or your belongings, but police may pat down your clothing if they suspect a weapon. Note that refusing consent may not stop the officer from carrying out the search against your will, but making a timely objection before or during the search can help preserve your rights in any later legal proceeding.

If you are arrested by police, you have the right to a government-appointed lawyer if you cannot afford one.

You do not have to answer questions about where you were born, whether you are a US citizen, or how you entered the country. (Separate rules apply at international borders and airports as well as for individuals on certain nonimmigrant visas, including tourists and business travelers.)

HOW TO REDUCE RISK TO YOURSELF: Stay calm. Don't run, resist, or obstruct the officers. Do not lie or give false documents. Keep your hands where the police can see them.

### I've been pulled over by the police

YOUR RIGHTS: Both drivers and passengers have the right to remain silent.

If you're a passenger, you can ask if you're free to leave. If yes, you may silently leave.

HOW TO REDUCE RISK TO YOURSELF: Stop the car in a safe place as quickly as possible.

Turn off the car, turn on the internal light, open the window part way, and place your hands on the wheel. If you're in the passenger seat, put your hands on the dashboard.

Upon request, show police your driver's license, registration, and proof of insurance.

Avoid making sudden movements, and keep your hands where the officer can see them.

### The police are at my door

YOUR RIGHTS AND HOW TO REDUCE RISK TO YOURSELF: You should not invite the officer into your house. Talk with the officers through the door and ask them to show you identification. You do not have to let them in unless they can show you a warrant signed by a judicial officer that lists your address as a place to be searched or that has your name on it as the subject of an arrest warrant.

Ask the officer to slip the warrant under the door or hold it up to the window so you can read it. A search warrant allows police to enter the address listed on the warrant, but officers can only search the areas and for the items listed. An arrest warrant has the name of the person to be arrested.

Even if officers have a warrant, you have the right to remain silent. You should not answer questions or speak to the officers while they are in your house conducting their search. Stand silently and observe what they do, where they go, and what they take. Write down everything you observed as soon as you can.

### I've been arrested by the police

YOUR RIGHTS: Say you wish to remain silent and ask for a lawyer immediately. Don't answer any questions or give any explanations or excuses. If you can't pay for a lawyer, you have the right to a free one. Don't say anything, sign anything, or make any decisions without a lawyer.

You have the right to make a local phone call. The police cannot listen if you call a lawyer. They can and often will listen to a call made to anyone else.

HOW TO REDUCE RISK TO YOURSELF: Do not resist arrest, even if you believe the arrest is unfair. Follow the officers' commands.

The bottom line is, if the cops are ever questioning you, you do not have to talk to them, no matter what they say. They might say something like: "The best thing for you to do right now is to just tell me the truth," or "Hey, buddy, I know you're a good guy, I get that, so just level with me and tell me what happened," but that is not the case.

The best thing for you to say: "Am I free to go?" And if you are free to go, then go. If you're not, then say, "I am exercising my right to remain silent, and I would like to speak to my attorney," and *nothing else.* If necessary, you can repeat this.

Do not try to fill any awkward silences with small talk. This is not a baby shower, and the consequences for saying something wrong in this situation are far more severe than Lauren gossiping about you to Heather at brunch over the weekend.

Also, never consent to a search. If they ask, simply say something along the lines of "Sorry, but my privacy is important to me, so no," or even simply "No, I do not consent to a search."

# ACKNOWLEDGMENTS

Thank you so much to Harper Collins for believing in this project. Huge thanks to my editor, Eric Nelson, and to my agents, Eve Attermann and Mark McGrath, for all of your help in making it possible to get this out there.

To everyone at Fox News! Big thanks to Suzanne Scott . . . who has worked so hard to give all of us at *Gutfeld!* this amazing platform and the freedom to share our views and make the jokes we want to make, especially in a culture that works so hard to censor that.

Thanks to Greg and Tom O'Connor for giving a little twenty-five-year old comedian, writer, and weirdo her shot; to my co-captain, Tyrus; and to everyone else on the *Gutfeld!* team.

To Cam, who knew from the start that I am not for the weak but chose me anyway. Thanks for always being understanding, even when I decided to dedicate my first book to a cat and a dead stranger instead of to my husband. I'm the best me because I'm with you.

To my family—my father Daniel aka "Dad Timpf"; my sister, Julia; my brother Elliott; and my brother from another mother, Keith. Your support and love are invaluable to me. Thanks to my dead mom, Anne Marie Ochab Timpf. I'd give anything to see you again, and I wish so much that you could see this.

To my French bulldog, Carl—for sitting on my lap while I wrote this, and knowing he's good-looking enough to not have to be jealous that I dedicated this book to Cheens and not him.

Thanks to everyone who read this book ahead of time and offered suggestions and/or advice, especially Elisha Maldonado, who is a brilliant editor and an even better friend.

Last but not least, thank you everyone who has ever let me crash on their couch along the way. I appreciate you, and am still sorry about the mess.

# INDEX

# ABOUT THE AUTHOR

KAT TIMPF is the cohost of *Gutfeld!* and a Fox News contributor. She has also worked at *National Review* and Barstool Sports and was a stand-up comedian long enough for this to be her third time quitting. You may recognize her from her more than ten years of studying and writing about speech, or from being the worst waitress you've ever had. She lives in New York City with her husband, Cam, their dog, Carl, and her cat, Cheens.